Smart Innovation, Systems and T

Volume 249

Series Editors

Robert J. Howlett, Bournemouth University and KES International, Shoreham-by-Sea, UK

Lakhmi C. Jain, KES International, Shoreham-by-Sea, UK

The Smart Innovation, Systems and Technologies book series encompasses the topics of knowledge, intelligence, innovation and sustainability. The aim of the series is to make available a platform for the publication of books on all aspects of single and multi-disciplinary research on these themes in order to make the latest results available in a readily-accessible form. Volumes on interdisciplinary research combining two or more of these areas is particularly sought.

The series covers systems and paradigms that employ knowledge and intelligence in a broad sense. Its scope is systems having embedded knowledge and intelligence, which may be applied to the solution of world problems in industry, the environment and the community. It also focusses on the knowledge-transfer methodologies and innovation strategies employed to make this happen effectively. The combination of intelligent systems tools and a broad range of applications introduces a need for a synergy of disciplines from science, technology, business and the humanities. The series will include conference proceedings, edited collections, monographs, handbooks, reference books, and other relevant types of book in areas of science and technology where smart systems and technologies can offer innovative solutions.

High quality content is an essential feature for all book proposals accepted for the series. It is expected that editors of all accepted volumes will ensure that contributions are subjected to an appropriate level of reviewing process and adhere to KES quality principles.

Indexed by SCOPUS, EI Compendex, INSPEC, WTI Frankfurt eG, zbMATH, Japanese Science and Technology Agency (JST), SCImago, DBLP.

All books published in the series are submitted for consideration in Web of Science.

More information about this series at https://link.springer.com/bookseries/8767

Óscar Mealha · Mihai Dascalu · Tania Di Mascio
Editors

Ludic, Co-design and Tools Supporting Smart Learning Ecosystems and Smart Education

Proceedings of the 6th International Conference on Smart Learning Ecosystems and Regional Development

 Springer

Editors
Óscar Mealha
Department of Communication
and Art/DigiMedia
University of Aveiro
Aveiro, Portugal

Mihai Dascalu ⓘ
Department of Computer Science
University Politehnica of Bucharest
Bucharest, Romania

Tania Di Mascio
Department of Engineering Computer
Science and Mathematics
University of L'Aquila
L'Aquila, Italy

ISSN 2190-3018   ISSN 2190-3026 (electronic)
Smart Innovation, Systems and Technologies
ISBN 978-981-16-3932-6   ISBN 978-981-16-3930-2 (eBook)
https://doi.org/10.1007/978-981-16-3930-2

This Springer imprint is published by the registered company Springer Nature Singapore Pte Ltd.
The registered company address is: 152 Beach Road, #21-01/04 Gateway East, Singapore 189721,
Singapore

Preface

It is our pleasure to announce another Springer edition of the International SLERD Conference Proceedings, an edition that still results from work and contributions that occurred during the global COVID-19 pandemic. This proceedings also confirms the resilience of our scientific community and the need to come forth with research results and foster solutions for the desired post-pandemic future. In 2020, the pandemic has affected the learning processes all over the world and learning ecosystems have reacted, exhibiting different degrees of resilience and promptness with the emergence of similarities, differences and inequalities. Multiple factors concur—i.e., processes, social communities and individuals—to determine the technological evolution of learning ecosystems and their smartness. The pandemic has generated a higher awareness about the intrinsic nature of the smart learning ecosystems which in return have evolved and have been influenced by the pandemic. The two dimensions that characterize the smart learning ecosystems—physical and virtual/digital (phygital)—are at stake, and a balance has to be re-achieved. Lessons learned with the compulsive global pandemic lockdown and the perpetuated online activity will surely start marking the agenda of a new era, an era with more certainty on the most adequate phygital models for learning ecosystems.

SLERD 2021 was announced with an expected venue in Bucharest, Romania, hosted by the University Politehnica of Bucharest; unfortunately, the global pandemic situation pushed, once again, this scientific event into a virtual fully blown, virtual, online international conference. The event was organized by the University Politehnica of Bucharest, Romania, with the partnership of Association for Smart Learning Ecosystems and Regional Development (ASLERD), an international nonprofit interdisciplinary and scientific–professional association committed to support learning ecosystems to get smarter and play a central role to regional development and social innovation. The University of Aveiro, Portugal, represented by its DigiMedia Research Group, also played a central role in the co-organization of this SLERD virtual edition.

Submissions received had authors from 13 different countries, an indicator of the global interest and partnership in the promoted topics. After a rigorous double-blind peer-review and meta-review process, 17 full papers were accepted to be included in SLERD 2021 Conference Proceedings, published by Springer, in the

Series *Smart Innovation, Systems and Technologies*. The selected scientific papers were organized by the conference topics and structured into four parts: (i) People in Place Centered Design for Smart Learning; (ii) Supportive Technologies and Tools for Smart Learning; (iii) Observing and Studying Learning Ecosystems; and (iv) Methods, Processes, and Communities. We believe this proceedings will be relevant for researchers, postgraduate students, teachers, designers and policy makers concerned with learning ecosystems, methods and empirical evidence of alternatives for post-pandemic learning scenarios.

It has been an honor to belong to this scientific community and serve as publishing chairs in this SLERD edition, with a final selection of papers that represent the excellence of all the authors' work, coupled with rigor throughout the decision and publishing processes. Much of this work would not have been possible without the effort and support of our Conference and Program Committees that included more than 50 international researchers. We would like to thank them all, for their time to organize the event with enthusiasm and commitment.

Aveiro, Portugal   Óscar Mealha
Bucharest, Romania   Mihai Dascalu
L'Aquila, Italy   Tania Di Mascio
May 2021

Contents

**Correction to: Robots as My Future Colleagues: Changing
Attitudes Toward Collaborative Robots by Means
of Experience-Based Workshops** C1
       Janika Leoste, Tõnu Viik, José San Martín López, Mihkel Kangur,
       Veiko Vunder, Yoan Mollard, Tiia Õun, Henri Tammo,
       and Kristian Paekivi

About the Editors

Óscar Mealha is Full Professor at the Department of Communication and Art, University of Aveiro (UA), Portugal. He develops his research at DigiMedia Research Centre/UA, in the area of "Information and Communication in Digital Platforms" in the context of "Knowledge Media and Connected Communities" with several projects, masters and doctoral supervisions and publications on interaction design and analysis techniques and methods, namely for UX design and evaluation, usability evaluation, and visualization of interaction/infocommunication activity. He is involved in infocommunication mediation projects such as "Unified Communication and Collaboration" with IT companies, "Visualization of Open Data Dashboards for Citizen Engagement and Learning" in municipalities and smart territories, and "Knowledge Interface School-Society (KISS)" with school clusters within the scientific network ASLERD. He is currently Director of the Doctoral Program on Information and Communication in Digital Platforms, a joint program of the University of Aveiro and University of Porto.

Mihai Dascalu is Full Professor at University Politehnica of Bucharest, responsible for the courses of Object-Oriented Programming, Semantic Web Applications, and Data Mining and Data Warehousing. He has extensive experience in national and international research projects with more than 200 published papers. Complementary to his competencies in NLP, technology-enhanced learning, and discourse analysis, Mihai holds a multitude of professional certifications and extensive experience on strategic projects on non-refundable funds (EU, WB, USTDA). Moreover, Mihai has received the distinction "IN TEMPORE OPPORTUNO" in 2013 as the most promising young researcher in UPB, has obtained a Senior Fulbright scholarship in 2015, has become a Fulbright Ambassador since 2018, and holds the US patent # 9734144 B2. Mihai is also Corresponding Member of the Academy of Romanian Scientists.

Tania Di Mascio is Assistant Professor at the Department of Engineering Computer Science and Mathematics of University of L'Aquila, Italy. She obtained a Ph.D. working on HCI research field, and she awarded a master's degree in Electronic Engineering. She cooperated with national and international research institutes. Her primary research activities are in HCI, user interface usability and accessibility, as well as in TEL, with a focus on information visualization and interaction paradigms. She was Coordinator of several European projects. She is Author or Co-author of more than 100 papers in peer-reviewed journals and international and national conferences. She is Member of Committees of international and national conferences. She is in the Editorial Board of International Journal on HCI and TEL research fields.

Part I
People in Place Centered Design for Smart Learning

Chapter 1
Smart Alternation Schemes and Design Practices During Pandemics

Carlo Giovannella

Abstract The pandemic outbreak has almost cancelled for a long period the possibility to use laboratories to carry on didactic activities, including those foreseen for the alternation schemes. In this paper, we report on an experience carried on in a vocational school of informatics, Ferrari school located in Rome, Italy. The alternation scheme conducted in "smart working" has been based on a design process (simulation of an innovation process) and has relied as much as possible on cloud applications to develop all phases of the collaborative process. Despite the lack of physical contacts, the experience has been largely appreciated by the students that had the opportunity to test the "smart working" modality. This case study and, overall, the outcomes of the survey demonstrate that the smart working is a viable opportunity for alternation scheme, that a design process can be conducted fully online with similar and even better results with respect to the face-to-face counterpart and, finally, that an online process offers the opportunity to better integrate the alternation scheme with the timing and organization of the high school curricula.

Keywords Alternation scheme · Student employability · Skill mismatch · Lifeskills · School smartness smart working · Innovation process · Incubator of projectuality

1.1 Introduction

The school could and should assume a renewed role of driving force in the creation and development of *smart territorial learning ecosystems* [1–4] capable to foster the integration of most of the actors operating in their territory of reference and, thus, the establishing of truly educating communities aimed at counteracting the students drop out, at taking care of the harmonious development of the individuals, at supporting their expectations and potentialities. Communities able, as well, to foster

C. Giovannella (✉)
University of Rome Tor Vergata – Dip. SPSF, Rome, Italy
e-mail: gvncrl00@uniroma2.it

ASLERD, Rome, Italy

© The Author(s), under exclusive license to Springer Nature Singapore Pte Ltd. 2022
Ó. Mealha et al. (eds.), *Ludic, Co-design and Tools Supporting Smart Learning Ecosystems and Smart Education*, Smart Innovation, Systems and Technologies 249,
https://doi.org/10.1007/978-981-16-3930-2_1

the integration of formal and informal learning and the contemporary development of an adequate set of skills: basic, transversal, digital, and vertical (these latter in particular, but not exclusively, in the vocational context).

In Italy, from one hand, such potential role for the schools is acknowledged and encouraged by the Ministry of Education that, especially in the pandemic context, is emboldening the schools to promote territorial pacts [5]. On the other hand, in the reality, schools are struggling to become a veritable pole of attraction, for example for companies that do not find sufficient reasons to invest, even in kinds (skills and human capital), into high and vocational schools and their students [6]. This criticality is even more evident in regions characterized by a low rate of manufacturing production and makes the job placement particularly complex. This especially for those students that from one side do not have the will or the possibility to enroll in tertiary education courses and, from the other, have not yet developed a sufficient bouquet of skills [7].

In this landscape, the alternation schemes assume a fundamental role because they offer the possibility to simulate working routines and foster the acquisition of knowledge, skills, and competences, complementary to the basic ones that are developed during standard curricular activities. Furthermore, in case the processes implemented as alternation schemes are capable to lead to valuable outcomes, these latter, once that have been adequately publicized, could help the schools to attract new actors and develop further the territorial network.

As already shown in the past [7], among the most interesting approaches to the realization of meaningful alternation schemes, we can include the *design processes*, either addressed to few selected students or, in a more massive form, to entire classes.

Unfortunately the pandemic outbreak has disrupted many alternation schemes since physical activities (including company internships) and laboratory sessions could no longer be realized. To cope with the consequences of various forms of lockdown, many schools have replaced practical activities with sessions of seminars to provide, at least, job orientation. In some others, as in the case that we will describe in this article, the design processes have been moved fully online to simulate a *smart working* configuration. A largely diffused adoption of the smart working, in fact, will be, likely, the most relevant heritage of the present pandemic, since in many cases it has no contraindications in terms of classical economic models. All companies (apart from the manufacturing activities), in fact, have realized that most of the activities can be transferred in the cloud, often with an increase of productivity and, for sure, with substantial savings (rental costs of offices/venues, utilities of various kinds, cleaning costs, meal vouchers, etc.) [8], also for the environment (reduction of the carbon footprint due to transportations and heating/air conditioning) [9]. Also the counterpart—the smart workers—seems to have appreciated the saving of time (especially commuters) and a better organization of personal time, provided that the working requests do not become too invasive. All these factors suggest the need for the students to experience possible future job modalities that, most likely, will be part of the so-called "new normal" [10]. Having this in mind, we have readapted the massive alternation scheme we implemented in the past—based on the simulation of an innovation process that has been named "Incubator of projectuality" (IP) [7]—in an online smart design activity. In the following, we first describe the organization of the "*smart alternation scheme*", then the outcomes of the process (deliverables) and those of a questionnaire that have been answered by the students that took part in the IP, together with the lesson learnt.

1.2 The Organization of the Smart Alternation Scheme

1.2.1 Process Design, Goals Sharing, and Starting Up

In a learning community, it is important to co-design all educational activities with the school's teachers and to share the goals of the process with the students (as main target) and, as well, with their parents to converge on the relevance of the alternation schemes for the future life of the adolescents. The co-design step has been performed during a videoconference session starting from the outcomes of the previous experiences and considering the peculiar situation generated by the pandemic. Accordingly, the design process has been divided in three main phases— problem setting, problem solving, and communication—keeping in mind also the three-layered *Organic Process* and the associated guidelines to carry on design activities [11]. For each activity we have defined the design methods to be used, the expected deliverables, its duration, the student's efforts, and the abilities/competences (LIFE [12] and digital skills [13]) the development of which is expected to be fostered. The co-design phase has also concerned the choice of the technologies to be integrated or used in parallel with the google classroom environment adopted by the Ferrari school. The presentation of the IP to both students and parents has been delivered by using Meet and has been followed by a Q&A session. Due to the online modality the participation of the parents has been higher than in the previous editions. The adhesion to the alternation scheme has been on a voluntary basis and in the case of the Ferrari school it was full: two classes (40 students) of the third year of the vocation school in informatics (15–16 years old). Only slightly more than half of them, however, completed the whole process, mainly due to competing interests and/or need to invest time to realign themselves with the standard curricular activities.

Following this first meeting, the students have been asked to fill an entry questionnaire on google form. The answers have been used as basis for an individual online interview (max 15') and to elaborate the Moreno diagrams [14]. The ensemble of the collected information has been used to form the design groups composed by 5–6 individuals, four groups per class. Groups' composition has been deemed optimal by about 95% of the students, at least at the beginning of the process.

1.2.2   Process Development

To develop the IP activities, all students and tutors have been assigned to a virtual class in the Google classroom environment of the school. We have been using only the stream as a dashboard to publish official communications and to link to the predefined Meet room that we used as common class. Beside the official environment, we used GDrive to create a common folder for the two classes and a sub-folder for each group of designers. Students have been invited to work collaboratively to produce, when needed, shared documents by means of GDocuments, GSheets, GForms and to upload in their sub-folder graphic elaborations and presentations. To carry on collaborative graphic work we adopted Mirò [15], while for low fidelity prototyping, we suggested the use of Balsamiq [16] and for medium fidelity prototypes that of Justinmind [17] or Figma [18]. Before beginning the design process, we provided the students with a time line, a list of the expected deliverables and the slides of the initial presentation.

All the meetings of the design process—12 meetings organized at least one each 10 days—have been composed by a theoretical pill (30'–45') delivered in the common Meet room and a practical and review session carried on in separate rooms, for a total of 2 h–2 h and half for each meeting. Theoretical pills dealt with design methodologies to be applied during the: (a) problem setting to include also data collections, data analysis, and data/problem representation; (b) problem solving; and (c) communication to include logo design, storytelling, and pitching.

The practical work started with a brainstorming to identify a design domain of interest and with the collaborative design of a conceptual map of such domain. To conduct parallel brainstorming sessions (each tutor has taken care of four sessions), we opened multiple rooms and used the Firefox browser because it allows very easily to mute/unmute each room and take part in one of the discussions while the others were going on (note that at that time breaking rooms were not yet available). Of course this technological setting has been possible due to the personal ultra-wide-band Internet connectivity of the tutors. Most likely, it could not have been actuated from school using, for example, a blended parallel configuration (50% of students attending the activity in the classroom and 50% of students participating in the activity from home) due to the limitation in the band connectivity suffered by many schools.

During the brainstorming, the students had also to design, in parallel, the concept maps on Mirò. A possible technical limitation of this choice is that students that connected to the working session by means of a smartphone could only participate in the brainstorming while other colleagues have to translate their suggestions on the Mirò board. On the other hand, such limitation offered a chance to train some of the LIFE skills the students were expected to develop.

Due to the limitations imposed to the free use of this collaborative design tool, we have divided each Mirò board into four areas and assigned each area to a different group (see Fig. 1.1). This strategy in the use of Mirò had several positive effects: the tutor could monitor at the same time the progress of four different groups and the students could have a look at the works of their peers. In fact, although it has not been

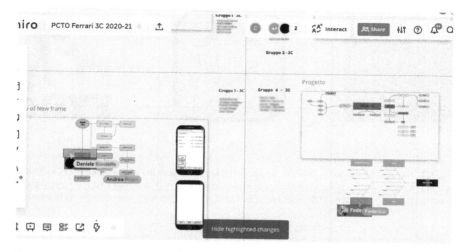

Fig. 1.1 An example of Mirò working session. The board has been subdivided in four working spaces. Daniele (yellow) and Andrea (green) are collaborating while Federico (red) is getting inspiration

possible to record the dynamics, it was possible to observe the displacement of the students around the boards, displacement that highlighted either the dynamics of the collaborative work going on inside the groups and the flights of students from one working area to another to get inspiration. It is worthwhile noting that our strategic approach to the use of Mirò, beside the possibility to uncover the working styles of the students, allowed also for behaviors that in the physical classroom would have never been possible to adopt because they would have generated chaotic situations and, thus, would have been forbidden. Each practical session of the process has been accompanied by a "home" follow up that was considered as a part of the students' effort.

The initial exploration of the domain and the efforts in the identification of possible problems/expectations, together with potential targets of the innovation, has been refined by: (a) a benchmarking on the Web to look for already existing solutions and/or inspirations; and by (b) the design of a questionnaire to be addressed to the potential target with the aim to verify the initial problem setting and better define the unavoidably ill-defined problems.

This phase, i.e. the problem setting, is usually—regardless of the modality of the process: online or face to face—the less attractive for the students and the most complex to realize because requires them a certain amount of efforts that are not immediately understood and that are not immediately paid back by striking results: "only" an optimal definition of the problem and few graphical representations! Not by chance the problem setting, although included in the so-called "design think-ing", is often skipped in many design-based educational processes that start directly with a briefing to provide a description of the problem and stimulate the immediate search for a solution. As far as the smart format of the alternation scheme, we have

Fig. 1.2 Examples of logos: **a** WePark, an app intended to support the social parking; **b** Cestinami, an app intended to support the separate waste collection; **c** FYDo (Find Your Dog) an app intended to support the identification of the most suitable faithful friend among those abandoned in a kennel

noticed that the large availability of cloud technologies allows for a much simpler organization of brainstorming, focus groups, interviews, design and delivery of questionnaires, etc., but all this is not sufficient to generate a substantially increase of the students' interest in the problem setting.

After the conclusion of the problem setting, the problem solving phase was initiated. However to keep high the level of involvement and stimulate, at the same time, a continuous refinement of the project—needed to minimize design errors and perform the feasibility and sustainability analysis—it was necessary to accompany the problem solving with the parallel development of the communication phase. After a rapid creative phase, in fact, students of this age tend immediately to start coding (even if they have not developed adequate skills) and try to get concrete results ("Hello world" effect). To increase the level of concreteness and immediacy, while keeping them on the refining of the project, we asked the students to design a logo, see Fig. 1.2 (some of which have been designed as smart logos, see Fig. 1.5) and find a name for their artifact.

To close the process, we asked the students to produce at least a low profile prototype (medium profile prototype for the most advanced projects, see Fig. 1.4), a storytelling as starting point to produce a short video clip and a pitch.

To keep high the momentum, we guaranteed a personalized (group by group) review sessions, at least each ten days, and keep updated a reviewing diary (google doc) with our remarks and advices on all deliverables under development. To stimulate the spirit of imitation, from time to time, we organized public reviews (to be considered as a sort of milestone check) during which the students have been encouraged to present to their peers the state of advancement of their projects. The overall design process took slightly more than 4 months and ended with a public online presentation (virtual pitching). Each group had full freedom in organizing and presenting the pitch which was given in a Meet room and, of course, took in consideration all the deliverables produced previously.

1.3 Outcomes

We measured the outcomes in two manners: on the basis of concrete deliverables and by mean of a survey based on Google form. Of course in this context, we cannot show all deliverables that have been produced by the students and that include questionnaires, benchmarking tables, data analysis and reports, perceptual maps, SWOT analysis, logos, definitions of missions and visions, *look&feel,* pitches, and in some cases preliminary medium profile prototypes and movies. Nevertheless in Fig. 1.2, we show some of the logos that have been realized by the working groups; in Fig. 1.3, we show a couple of examples of *look&feel* realized to develop medium fidelity prototypes of apps, and, finally, in Fig. 1.4, an example of a preliminary navigation tree.

It is important to underline once more that the students, before their participation in the IP, had only preliminary notions of HTML and CSS and have never tested themself in a design/innovation project and in the realization of any kind of prototype.

We would also like to stress that in a couple of cases, the students were able to elaborate dynamic logos capable of transmitting quick information to the user. For example, in Fig. 1.5, it is shown a sketch of a logo studied for an application dedicated to bus users that aims at informing the users on the number of next incoming bus, on the bus filling level (red filling) and on bus proximity (bus dimension).

The survey has been realized by means of a Google form that have been filled by slightly more than 50% of the participants. This percentage is smaller than the one that can be achieved in presence, when a paper questionnaire is distributed to all students, but still relevant.

The first and very relevant outcome concerns the level of the students' satisfaction, 8.05/10 on average, much larger than the one detected in the previous edition

Fig. 1.3 Example of *Look&Feel*: **a** Cestinami (already described in figure caption 2); **b** School Map: an app intended to support the student in finding the right "path" to follow within the school

a)

b)

Fig. 1.4 Example of the navigation tree of the app Cestinami (medium profile prototype)

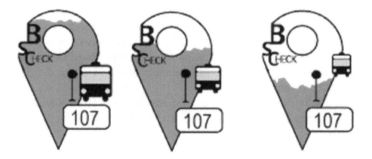

Fig. 1.5 Example of "smart" (dynamic) logo designed for an app intended to support bus users

conducted in presence [7]. In Fig. 1.6, we show the distribution of the students' satisfaction measured over several years either for all the alternation schemes realized by the school for the students of the III years and in the case of the incubator of projectuality, IP.

The increment of the students' satisfaction during the last "*smart edition*" appears very clear (mode of the distribution equal to 8—yellow line), as it is clear that the displacement of the distribution toward higher value cannot be affected by the relatively low percentage of the respondents. In fact the lower participation could influence at most the average value of the satisfaction (due to the lack of contribution for values lower than six) but not the higher prevalence of satisfaction, values around 8 and 9 with respect to all previous processes. It is also interesting to note that the average level of satisfaction measured for the massive *smart alternation*

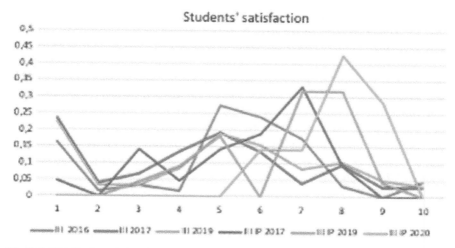

Fig. 1.6 Distribution of the III year students' satisfaction for the alternation schemes organized by the vocational school in informatics (Ferrari–Rome). The measures have been carried out for several years: 2016, 2017, and 2019 for the students of all III classes of the school; 2017, 2019, and 2020 for the students that took part in the incubator of projectuality (IP), i.e., the alternation scheme described in the present article

scheme described in this article is as high as those measured for other processes that involved smaller and selected groups of motivated students [7].

As far as the activities conducted during the process here, the main outcomes of the survey:

(a) the theoretical pills should be no longer than 30 min with an optimal duration around 20 min;

(b) 95% of the students stated to have fully understood the relevance of all phases of the process they went through;

(c) the most appealing topics, apart the introduction to the innovation processes (that made a large use of practical examples and videos), are those that stimulated the students' creativity without demanding too much effort in learning tools and methodologies, like the definition of the logo and the search for the name of the application (around 75% of popularity), the definition of the app functionalities and the design of the low fidelity prototype (around 50% of popularity). Less popular are the most demanding and conceptual activities like: storytelling, storyboarding and video making (30%), pitch design and realization (30%), medium profile prototyping (25%), and the definition of the vision and mission of a potential start-up (25%);

(d) overall the students declared to have developed a consistent set of LIFE skills, in particular: collaboration and team working (63%), decision making (63%), process management to include resources management (58%), self-regulation (53%), proactivity (53%), flexibility and adaptation (47%);

(e)   the level and typology of the challenges presented by the performed activities
        have been appreciated, 7.39/10; students affirm that *"the process fostered them
        to collaborate on project ideated by themselves; also with peers they have never
        collaborate with before"*; *"they learnt to respect the deadlines"*, *"they learnt
        to work in parallel on many different aspects, most of which were completely
        new for them"*, *"they had to accept remarks and work on them to improve their
        project"*

(f)   the students' motivations mostly satisfied by the process have been: (i) the
        acquisition of abilities and competence; (ii) the growth of their cultural back-
        ground; (iii) the possibility to discuss and compare opinions, and (iv) the
        stimulus to the creativity;

(g)   the orientation action for the future has been considered positive, 7.42/10,
        because: *"made me consider more opportunities for the future"*, *"I had to use
        applications that could be useful in the future, i.e., video/photo editors"*, *"I had
        the possibility to experience team working and team management"*;

(h)   the contemporary presence of more than one tutor has been largely appreciated,
        7.79/10, basically because they could provide different and complementary
        points of view on the deliverables during the revisions.

The opinion of the students was rather positive also as far as the quality of the
online delivery of the process and its organization, 8,00/10: *"Despite the current situation, the organizers were more than skilled in solving the problems and proceeding
with the activities"*; *"they provide us all that was needed and allow us to move forward
with our times"*; *"they manage to get us into a work perspective that is very different
from the scholastic one learned up to now"*.

It is interesting to note that the students claim to have been able to organize their
time very well: + 2.77 on a scale ranging from −5 to + 5.

Of course students fully recognize the role and the relevance of the technologies,
8.42/10: *"Without the technologies during the lockdown it would have not been
possible to carry on similar activities"*; *"all the documents and the deliverables
have been created using computers, smartphones or tablets"*; *"with technologies it is
possible to do everything, and the technologies in this alternation scheme has been
very well used"*.

Nevertheless 90% of them do not deem necessary to use the camera in online
interaction; they believe that voice contact and screen sharing are sufficient to discuss
and analyze the on-going work.

Despite the above very positive judgments when the students have been asked if
the alternation scheme can be carried out completely online, the average value of the
numerical answers, 7.16/10, although not at all low, was not as high as those detected
for other measured parameters. This outcome could be explained by the students'
wish to spend time in as many as possible companies with the hope to initiate the
development of their own personal system of relationships that, of course, requires
more intimate contacts [7].

Another interesting outcome is that about 71% of students are available to carry
on this type of activity even in the afternoons and online. This suggests that it is a type

of activity that, for the future, could be easily organized in a blended configuration without a relevant impact on the usual curricular courses and activities.

1.4 Lesson Learnt

The experience reported in this article shows very clearly that, at least in all contexts for which innovation processes are relevant, the *smart alternation schemes* constitute an interesting alternative to standard alternation schemes realized in presence, whatever their nature [7]. The pandemic has probably cleared the *smart working,* and it is therefore more than ever very useful for students to experience alternative modality of work, that in a close future might become their natural modality to work. Alternation schemes based on smart work, furthermore, may favor a larger participation of companies in alternation schemes since on the logistical point of view it would be less expensive for them.

The evidence gathered during this case study shows that students can be very good at smart working, very clever in mastering cloud collaborative working environments, and in using them to carry on complex workflows like those that characterize innovation-oriented and design-based processes. Furthermore, the use of such environments seems also to allow for the overcoming of possible technological gaps among students, thanks to the reciprocal support and a division of tasks also driven by the technologies and the devices available to each one of them.

Students need to be exposed to alternation schemes such as the one described in this work because the school does not offer them the opportunity to work on complex projects, of medium-long duration and in which they are obliged to optimize the distribution of tasks and, as well, to actuate a non-scholastic cooperation necessary to achieve the final goal.

Once the students have gone through all the phases of a design process and experienced all the foreseen activities and methodologies, they are able to recognize their relevance, but to get to this point it is necessary to maintain a not simple balance between the provision of theoretical pills and students involvement in practical activities, between more conceptual and practical activities capable to generate concrete and immediate results, between freedom of action at the cost of achieving less than optimal results and constrains in order to avoid irreparable errors. Careful planning and adequate sharing of objectives can certainly help, but the right alchemy to keep students in a state of flow [19] must be built day by day, despite the fact that alternation schemes are time-deprived educational processes that require to condense all activities in few hours and which always leaves the students in a kind of suspension: *"I would like to see what was just an idea become concrete"*; *"I would like to continue because the work we have done is intriguing"*; *"I would like to continue to produce something concrete, functional and useful to anyone"*; *"I would like my school to use the app we designed"*.

Acknowledgements The author is particularly grateful to the School Principal, Prof. Ida Crea, for trusting in the alternation scheme described in this article and, as well, in its transposition in the "smart work" modality. The author is very grateful also to: Prof. Giuseppe Brandinelli, Bianca Ielpo, and Maria Cristina Solenghi for both sharing the design of the process and for acting as internal tutors; to Dr. Vincenzo Baraniello and Ing. Paolo Mondini for acting as external tutors; to all students of the III year of the vocational course in Informatics.

References

1. Giovannella, C.: "Smartness" as Complex Emergent Property of a Process. The Case of Learning Eco-systems, pp. 1–5. ICWOAL, IEEE Publisher (2014)
2. Hannon, V., Thomas, L., Ward, S., Beresford, T.: Local Learning Ecosystems: Emerging Models. WISE. https://www.wise-qatar.org/2019-wise-research-learning-ecosystems-innovation-unit/
3. Giovannella, C.: Participatory bottom-up self-evaluation of schools' smartness: an Italian case study. IxD&A J. **31**, 9–18 (2016)
4. Giovannella, C.: Schools as driver of social innovation and territorial development: a systemic and design based approach. IJDLDC **6**(4), 64–74 (2016)
5. MIUR: Idee e proposte per una scuola che guarda al futuro, p. 32. MIUR https://www.miur.gov.it/documents/20182/0/RAPPORTO+FINALE+13+LUGLIO+2020.pdf/c8c85269-3d1f-9599-141c-298aa0e38338?version=1.0&t=1613234480541
6. Giovannella, C.: Incubator of projectuality: an innovation work-based approach to mitigate criticalities of the Italian massive alternance scheme for the school-based educational system. IJDLDC **8**(3), 55–66 (2017)
7. Giovannella, C.: An analysis of alternation schemes to increase student employability and the smartness of secondary schools. In Ludic, Co-design and Tools Supporting Smart Learning Ecosystems and Smart Education, pp. 39–51. Springer Publisher (2021)
8. Vyas, L., Butakhieo, N.: The Impact of Working from Home During COVID-19 on Work and Life Domains: An Exploratory Study on Hong Kong. https://doi.org/10.1080/25741292.2020.1863560
9. Hook, A., Court, V., Sovakool, B.K., Sorell, S.: A systematic review of the energy and climate impacts of teleworking. Environ. Res. Lett. **15** (2020). https://doi.org/10.1088/1748-9326/ab8a84
10. Anderson, J., Rainie, L., Vogels, E.A.: Experts Say the 'New Normal' in 2025 Will Be Far More Tech-Driven, Presenting More Big Challenges. https://www.pewresearch.org/internet/2021/02/18/experts-say-the-new-normal-in-2025-will-be-far-more-tech-driven-presenting-more-big-challenges/
11. Giovannella, C.: An organic process for the organic era of the interaction. In Silva, P.A., Dix, A., Joaquim Jorge, A. (eds.) HCI Educators 2007: Creativity3: Experiencing to Educate and Design, pp. 129–133, 10 (2007). http://disco-tools.eu/disco2_portal/terms.php
12. http://disco-tools.eu/disco2_portal/terms.php
13. Carretero Gomez, S., Vuorikari, R., Punie, Y.: DigComp 2.1: The Digital Competence Framework for Citizens with Eight Proficiency Levels and Examples of Use. Publications Office of the European Union (2017)
14. Moreno, J.L.: Who Shall Survive? Beacon House, New York (1934)
15. https://miro.com/
16. https://balsamiq.com/
17. https://www.justinmind.com/
18. https://www.figma.com/
19. Czisikszentmihalyi, M.: Flow—The Psychology of Optimal Experience. Harper & Row (1990)

Chapter 2
Supporting Urban Innovators' Reflective Practice

Alberto Magni⑩**, Alicia Calderón González, and Ingrid Mulder**⑩

Abstract Over the past years, a growing number of local initiatives are generating solutions for societal challenges in their cities. However, the scale and complexity of these challenges force urban innovators to constantly adapt and learn, having to acquire new capabilities that will help them advance towards systemic change. In the current work, we take the premise that these urban innovators need to be able to utilise the urban context as a learning ecosystem in order to push their interventions beyond the boundaries of small innovative niches. In keeping with Schön's reflective practice, we envisage reflection as a core competence for these urban change makers to grow and present a reflective process supporting urban innovators in framing their professional learning journey to succeed in their projects. A series of online sessions have been conducted to investigate how to scaffold a reflective process enabling innovators to better identify challenges in their projects and the corresponding capabilities they need to acquire. In the proposed paper, we present reflective activities as a tool supporting urban innovators in self-defining their learning journeys and elaborate on the insights gained. It can be concluded that the reflective process we developed was valuable to urban innovators in unveiling new learning needs for their projects, while further research is needed to more effectively translate these learnings into actionable steps to sustain innovators' self-development.

Keywords Design · Learning ecosystems · Reflection · Self-development · Societal challenges · Urban innovators

A. Magni (✉) · A. Calderón González · I. Mulder
Delft University of Technology, Landbergstraat 15, 2628 Delft, CE, Netherlands
e-mail: A.Magni@tudelft.nl

A. Calderón González
e-mail: A.CalderonGonzalez@tudelft.nl

I. Mulder
e-mail: i.j.mulder@tudelft.nl

© The Author(s), under exclusive license to Springer Nature Singapore Pte Ltd. 2022
Ó. Mealha et al. (eds.), *Ludic, Co-design and Tools Supporting Smart Learning Ecosystems and Smart Education*, Smart Innovation, Systems and Technologies 249,
https://doi.org/10.1007/978-981-16-3930-2_2

2.1   Introduction

Societal challenges are increasingly spurring the emergence of a growing number of local initiatives that leverage the resourceful and interconnected nature of cities and use them as urban learning ecosystems to engage in and experiment with innovative and creative ways to generate social innovation. Interestingly, mature urban initiatives show oftentimes a diverse mix of backgrounds within their team and closer collaborators; think of designers, local authorities, academia, private and public organisations, who share an interest in proposing positive change. Design skills and approaches seem to be a promising resource for urban innovators, but also new capabilities are needed (e.g., community building, business acumen, strategic leadership, to name a few). More importantly, when implementing innovative solutions for societal transformations, these teams of urban innovators find themselves learning and operating in a complex, multilevel system that is the urban context and society as a whole. "Imagining, creating and developing these innovations requires the simultaneous consideration of different perspectives" [1], and the creation of propositions valuable for multiple actors at different levels; from citizens to private and public organisations, including local authorities. Urban innovation processes become then "co-evolutive" processes, where innovators must be capable of constantly learning from and with the different dynamics, actors, resources and competences that characterise the urban ecosystem [2]. This, in order to identify the most appropriate strategies and capabilities that can help them advance in embedding their projects in cities urban innovators need to continuously identify the most appropriate innovation strategies and capabilities [2]. To exploit cities as learning ecosystems, our premise is that innovators must be increasingly aware of the capabilities they need to embed innovation in cities, as well as the steps to acquire them. In keeping with Schön's reflective practice [3], the next section elaborates upon reflection as a core capability enabling these urban change makers to grow. Following a review on learning and reflection, we motivate a reflective approach enabling urban innovators to better frame their needed learning journey to succeed in their projects. A series of reflective design interventions have been conducted to investigate how to structure such a reflective process in a way that would help innovators identify the challenges ahead in their initiatives and the capabilities they need to acquire to achieve them. We report the designed interventions and the insights gathered on the structuring and support of a reflective activity for innovators. Afterwards, we elaborate on the insights gained and discuss the value of introducing reflective activities as a tool for urban innovators' development. We end with a series of guidelines for enhancing reflective practices that support urban innovators' ability to self-define their learning journey.

2.2 Learning and Reflection

Seen as an "active and purposeful process of exploration and discovery", reflection can become a promising tool for self-development [4, p. 496]. Particularly, as it helps to "become receptive to alternative ways of reasoning and behaving", reflection is likely to open up learning opportunities. Specifically, engaging in a reflective process can help practitioners to "improve (their) ongoing practice, by using the information and knowledge that they are gaining from experience" [5, p. 16]. Schon's distinction in reflection on/in action is particularly relevant to innovation practice; the ability to learn from experiences to frame how to better act on current unknown challenges. It can be concluded that through a "reflective conversation with the situation" at stake, innovators explore the problem situation relating it to past experiences that could help them approach it. In this way innovators can "name the things to which [they] will attend and frame the context in which [they] will attend to them", and by doing so they more easily define the initial problem, and together with it also the decision to be made, the ends to be achieved, the means which may be chosen" [3, p. 40]. Similarly, Dorst and Cross [6] argue that through a process of exploration of problem and solution spaces, designers can identify what are more specific, unresolved problems to focus on, on which they can focus their creative effort in elaborating new approaches. Designers explore the so-called problem and solution spaces, analysing first what are the elements of a given problem situation, to then relate them with elements of previous situations they have already encountered. This helps them identify initial solutions they know to approach the situation, which are, however, likely to solve the problem at stake only partially. By elaborating partial solutions, designers can isolate and frame what are new, more latent, and more specific challenges from the initial problem situation they had. Challenges that they are still unable to solve and for which they need to frame and experiment with new creative approaches and solutions. In the current work, we refer to a similar reflective process as a key element in urban innovation initiatives' ability to learn. More specifically, we address social innovators' necessity to constantly learn and adapt throughout the complex process of urban innovation, to identify the needed capabilities to be developed to succeed in their projects with social impact [2]. In keeping with leading scholars [3, 6], if urban innovators systematically engaged in a reflective process on their projects, they would likely better define what are the new specific requirements of the different tasks involved in their projects that would force them to develop new capabilities. This potential, however, remains in theory. While reflective approaches have been successfully used for professional self-development [3–5], it is not straightforward how urban innovators can benefit from these approaches when dealing with complex multifaceted challenges [7]. For these reasons, we set out a study to investigate how reflection could practically become a tool for innovators to identify the challenges ahead in their initiatives and the capabilities they need to acquire to achieve them. The following section introduces the context, the approach taken, and the methodological details of the study.

2.3   Study

The current work is part of the European DESIGNSCAPES programme that investigates how to ignite the transformative power of design for sustainable and responsible innovation in European cities. The programme supports a hundred mission-driven urban innovation initiatives, tackling complex societal and environmental issues connected to the European sustainable development goals. More specifically, these initiatives are selected at three different stages of their innovation process: when carrying out their initial feasibility studies, when prototyping and embedding their solutions in an urban context, and finally in their stage of replication to a new city. Next to funding, the programme supports these initiatives with a training programme that has been developed with a twofold goal. On the one hand, to identify which capabilities are key to foster social innovation and urban development, and, on the other, to provide appropriate training to infrastructure a community of innovators that can continue learning and developing in a self-sustaining way, beyond the programme itself. Within this context, the present study investigates how to facilitate the growth and self-development of urban innovation initiatives; introducing a reflective approach enabling urban innovators to better frame the challenges ahead in their initiatives and the necessary professional learning needs. Participants have been recruited from the awarded initiatives in the European programme. Initiatives that were active in prototyping their proposed urban innovation and were engaged in a complex process of embedding innovation in urban contexts were invited ($n = 30$). Out of the 30 invited initiatives, fitting that criteria, seven initiatives participated in the study. In the first two sessions, only the contact persons participated, and from the third session onwards it was explicitly asked to participate as a team, accumulating a total of 15 urban innovators participating in our research activities. The initiatives varied in the kinds of urban challenges addressed, which ranged from the revitalisation of communities and urban spaces through active participation in co-design processes to the use of platforms to facilitate the promotion of more sustainable mobility behaviours as well as the increase of public awareness on noise pollution in cities. The initiatives took place in seven different medium and large cities located in four European countries: Italy ($n = 3$), Bulgaria ($n = 2$), Greece ($n = 1$), The Netherlands ($n = 1$). Moreover, the background and composition of the different initiatives' teams varied, including among others, practitioners from architecture, design, software development, cultural heritage, and environmental engineering. Table 2.1 summarises the respective project goals and locations of each initiative participating in the study.

Innovators' reflections on their projects are not expected to be immediately noticeable, but rather a mental process carried out internally. With this in consideration, our research approach entailed a series of five interventions where we used design artefacts to trigger innovators' reflections. The insights of each intervention have informed the design of the following one, to gradually develop a process for a reflective activity. In particular, these interventions have been designed as reflective sessions to help urban innovators think of the future steps of their projects, compare them with their previous experiences in practice, and in this way identify what are new challenging aspects in their current projects that force them to develop new capabilities and skills. Next to developing a reflective process, we also have gathered insights on the design requirements of the supporting material of this process. Figure 2.1 shows the examples of the material used in the first and the last

Table 2.1 Project goals and locations of participating initiatives

Initiative	Project goal	Country
Team 1 ($n = 3$)	Connects local institutions, architectural heritage owners, residents, artists, architects, and investors to revitalise disappearing city centres	Bulgaria
Team 2 ($n = 1$)	Promotes grassroots transformation and reuse of existing urban spaces, through a platform enabling citizens to identify sites, generating proposals, and access a variety of funding options	Bulgaria
Team 3 ($n = 1$)	Co-creation with citizens of a network of urban landmarks (urban art structures) linked to existing green spaces in the city, aimed at the reactivation of neighbourhoods' public spaces and social connections	Greece
Team 4 ($n = 3$)	Mapping noise pollution in urban environments, raising public awareness and helping people suffering from hearing impairments to better cope with loud city areas	Italy
Team 5 ($n = 2$)	Fostering community resilience to climate change through co-design of green–blue infrastructuring in urban parks	Italy
Team 6 ($n = 4$)	Develops a service for local administrations offering a co-design instrument to involve citizens in the planning of public/common spaces using temporary architectural devices,	Italy
Team 7 ($n = 1$)	Developing a service for employees and companies to find sustainable mobility solutions tailored to local needs to encourage sustainable commuting behaviours	The Netherlands

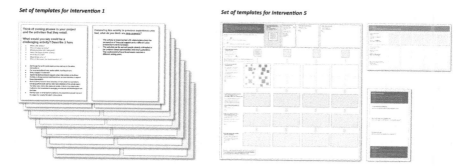

Fig. 2.1 Supporting material used in the reflective sessions. Template used for the first intervention (left) and the final template set informed by the first four interventions (right)

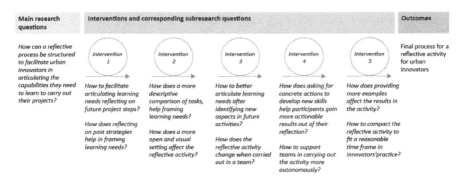

Fig. 2.2 Overview of the research approach and research questions for each intervention

intervention. Figure 2.2 shows an overview of our research approach and the five interventions with respective research questions.

Initiatives' members were invited via email to take part in online reflective sessions, which were introduced to them as research activities from the European programme. The first intervention was carried out twice, each time with a different initiative, to compare results. The same has been done for the second and third interventions. In the fourth and fifth interventions, only a single initiative took part. Five different sets of templates that helped, one for each intervention, were designed to structure the reflective process that innovators would go through. Participants' interaction with the digital materials developed allowed for observing their reactions and behaviours in the activity, collecting insights on their reflective process as well as on design requirements to develop more supportive templates. Online tools such as Skype or Zoom were used to communicate with participants, while collaborative tools such as Google Slides and Miro were chosen to structure and host the designed artefacts and carry out the interactive activity online with participants. The online reflective sessions have been audio and video recorded. Additional data has been collected via notes taken during observations of the participants' behaviours during the session. Next to that, at the end of each session, feedback interviews have been conducted to gather the participant's impressions on the interventions. The next section reports on the results from our approach.

2.4   Results

In the following section, we describe the five resulting design interventions as well as the corresponding insights of each intervention. The first intervention has been designed as a slideshow containing reflective questions that have been discussed together with urban innovators. The innovators participating in this first intervention were asked to think of a challenging future activity in their projects, and the aspects in this task that may pose new challenges. Afterwards, participants were asked to think of strategies they used in the past to solve one of the named challenges and elaborate on how they would solve it now, and what would be remaining issues. From this intervention, we observed the need to provide an overview throughout the entire reflection process to the participants, so that they could look back to their answers during the activity. We observed that when participants described more nuanced aspects of the tasks at hand, this helped them define specific "challenges within the challenges", that seemed promising directions to identify their learning needs. On the contrary, when participants reflected on how to apply past strategies to their current tasks, they stopped exploring the challenging situation at hand and focused instead on elaborating solutions to solve it, which was inherently complicated.

Informed by the first intervention, the design of the second intervention includes a more open structure, with templates designed on a digital whiteboard on the collaborative online tool Miro. Participants were asked to choose one of the future activities in their project to focus on, divide this into steps, to then reflect on one of these alone. They were presented with prompt questions to help them elaborate on who and what will be entailed in the chosen step, to contextualise it. These same questions were posed to describe a similar past activity. After articulating differences and similarities between the two activities, participants were asked to list the challenges they perceived in the new task and, in consequence, what learning they expected to need to face them. Insights on the reflective process confirmed that comparing the current and past tasks pointing to specific contextual aspects (e.g., differences in terms of stakeholders, location, methods) helps nuance specific aspects to tackle and learn for. However, a too detailed description of the tasks also slowed down the activity, interrupting the participants' rhythm. The open structure made the activity slightly scattered, requiring consistent facilitation from the researcher to navigate through the digital board.

For the third intervention, a set of three templates was designed to propose a more structured process, to help participants better reflect autonomously. The process of selecting a step to reflect on and initially describing it was designed to be shorter, while more focus was put on asking participants to elaborate on the skills they showed in past similar experiences. Gibbs' reflective cycle [8] was used as inspiration to guide participants in describing and evaluating past experiences through prompting questions from the steps "description", "feelings" and "evaluation" of his framework. Consequently, we also created more space for the participants to elaborate on what differences in the current task required them to adapt their capabilities or learn new ones. We observed that without elaborating first on the chosen step, participants reflected on the skills required too broadly. When asked "Based on this, which capabilities do you think you possess?" and "What skills could be required to carry out these tasks successfully?" participants did not explain to what extent their capabilities would be effective in a given situation. Rather, participants tended to abstract capabilities from the situation to tackle. However, carrying out the activity as a team helped the participants build on each other's answers, offering different points of view on the experiences and tasks at hand, leading ultimately to more nuanced answers. Moreover, it also benefited the teams' internal alignment, as some team members were not updated on specific plans, and the reflective activity helped not only bring those up to speed but also bring the team together to further detail these plans. Another observation was that asking participants at the end of the activity questions like "Where would you go to learn this? Who could you ask?" helped them formulate a plan of action to acquire the now needed skills. We also understood the need of reducing written tasks in the activity to leave more room for discussion among team members, as the most interesting reflections happened when the participants interacted with each other. Finally, it was observed that participants were not able to carry out the reflective activity by themselves, and facilitation was required to guide them through the different steps presented in the templates.

In the fourth intervention, the reflective process has been adjusted to first ask participants to define what the task selected requires and the capabilities they thought would help them achieve it. To avoid the formulation of capabilities in abstract terms, observed in the previous intervention, this time participants were not asked "What capability would be necessary?", but "What do you need to be good at, to achieve this task?" instead. After that, they were asked to recall past experiences in which those skills had already been used, to compare those situations with the present task, and consider if those skills needed to be adapted. In this intervention participants also had to answer questions about which concrete actions they would take to acquire the new capabilities needed, and a fourth template was added to write down takeaways from the activity as well as the team's next steps articulated. Instructions for participants on how to move from one template to the next have been improved to make it easier for them to follow the activity autonomously. In this fourth intervention, it was observed that expressing what capabilities a task requires was not easy for participants, instead it created ambiguity on what the word "capability" stood for. Asking instead "what do you need to be good at", was received more positively by participants, as the question opened up to multiple possible answers and allowed for giving more detail than just voicing specific capabilities. For example, one participant said, "It may be hard to express immediately as a capability the fact of being able to engage experts". Another observation was that participants found the added template for noting down takeaways a valuable instrument to summarise what they learned throughout the activity. The questions on actionable steps "What are concrete actions you can take?" generated interesting answers from participants that included approaches and strategies to acquire the missing capabilities (e.g., in this case, the group was missing "scientific capabilities", and identified as strategy the structuring of collaborations with researchers for publications). Overall, the participants could carry out the activity almost autonomously, however, guidance was still needed when going from one template to the next one and the activity took too long for a session that the team would carry out in their practice.

The insights gained thus far informed the design of the fifth and last intervention and corresponding reflective activity, which has been structured into three main sections, with a different template per section. In the first section, participants list and choose a future task in the project that the team wants to reflect on. Supported by prompting questions, in the second section, participants described first what the project task entailed and what they expected they would need to do. Secondly, they articulated what they needed to be good at to succeed in the task, listing the main capabilities that are required in their opinion. Participants were then asked to recall and describe some (up to four) previous experiences in which they already used one of the required capabilities, and then compare those situations with the circumstances of the current task, to uncover new challenging aspects for them in this instance. Once these new aspects were found, innovators are asked to define the new skills that they think they need to develop if they want to succeed, and the steps they could take to develop them. The third and last template helped participants to summarise the main learnings and next steps that have been mapped throughout the session.

The fifth intervention provided additional insights regarding both the structuring of the reflective process and requirements for supporting material. It was noticed how asking to recall and describe multiple examples made the activity slightly longer with the risk of going off track if there was no intervention of a facilitator, (role played by the team leader of the project in this instance). However, multiple examples also helped to further define differences between current and past situations, resulting in better answers for the reflection. When asked "What capabilities do you need to develop?" participants more often replied with new requirements or new tasks they had to be capable of doing (e.g., "to mediate now is fundamental to share the knowledge with the people there"). Sometimes with really detailed answers, such as "we need to interpret quickly the inputs we will get, without waiting too much we need to concretise what they tell us somehow". Such rich information, however, was then hard to express concisely as a capability, as the question suggested, and some information was therefore lost in the outcomes of the reflection. Defining ways to acquire the emerging needed skills was difficult for participants, especially when these were new to them. When the answers given as strategies for the next steps were too generic, participants considered them banal and not actionable enough. However, it was also observed that when participants were able to think of professional profiles from whom to learn, their answers seemed promising as a base to take further action towards fulfilling their learning needs. Overall, the templates showed to be clear and self-explanatory, since participants could carry out the entire activity without the need for external facilitation. Nevertheless, the whole procedure still needed streamlining and the number of questions and steps could be reduced. The time needed to conduct the session remained a limiting factor, for which also the presence of a facilitator was crucial. Finally, the innovators' feedback on the structure of the activity points to the need of having a more flexible and modular setup that could adapt to their time available or need to reflect on multiple steps of their project instead of only one.

2.5   Discussion

In the reminder, we discuss the main steps distilled from our learnings that seem to be helpful in a reflective process enabling urban innovators to identify learning needs and provide accompanying guidelines for better supporting reflection.

One observation was that reflection is not straightforward and the participants need to be guided to reflect and define their present challenges. We first helped participants to detail what they want to obtain, and what they would imagine the task will involve, to uncover and highlight what will be required of them. Articulating the requirements of the task as a first step, seemed to ensure that the participants' reflection revolves around the relevant aspects that will determine the success of the task. We continued by asking participants what they think they need to be good at to achieve the task at hand. As seen in the last two interventions, open questions encourage spontaneous and nuanced answers and do not restrict participants to think only in terms of skills or competencies. More interestingly, participants provided actionable tasks to complete instead. To reflect on what they are already capable of doing was a helpful stepping stone for participants; for example, by recalling past experiences and analysing what actions they took and what resulted from them, and in what circumstances they acted. This reflection mechanism helps first clarify the challenging aspects of new tasks to consider, to then let participants reflect on how these new aspects force them to develop further their capabilities; articulating once again what is that they need to be capable of, or good at, to succeed in their new tasks. These new required actions will likely point to the new capabilities that innovators need to acquire and help them formulate the initial steps to continue their learning process.

Through our study, we also identified a series of aspects to consider when setting up and facilitating participants' reflection on learning needs. For example, to carry out a reflective process following the steps previously presented, especially with a team, the role of the process facilitator is key. Another important lesson is to avoid including excessive steps and instructions when structuring reflection, as this can hinder the reflective and discursive nature of the activity. It seems to be constructive to create enough space for the participant's personal "flow" of reflection to lead part of the process. Another important aspect is to guide innovators to contextualise the capabilities they may have exercised in the past and avoid discussing them as abstract qualities. As different circumstances change our ability to achieve the same task, being capable of doing something, therefore, is directly related to the specific context in which that is carried out. As in Schon's reflective conversation [3], here it is valuable to guide participants to recall the elements of a situation that one had attended or will attend, and frame that situation reflecting on what circumstances make (or made) them capable (or not) of achieving something with the skills they already possess, and how this can play in the new task or challenge at hand. To better unveil these specific circumstances of a current task, it might be best to compare the latter to multiple examples of related previous experiences and their respective aspects or situations (e.g., recalling who was involved, what was the intention). In this sense, in line with Gibbs's reflective cycle [8], which has proved to be a valuable inspiration in supporting reflection on prior experiences, asking participants to provide examples, episodes, or even anecdotes, contributes to a more effective comparison and reflection on the concrete different aspects of the task at hand. The sessions also highlighted that reflecting in a team can provide additional benefits, compared to individual reflections. This happens especially if the team involved can discuss a shared experience and build on each other's answers, refining better the problem and solution spaces, and improving significantly the framing of the newly uncovered challenges.

Whereas insights are (co)-constructed during discursive and reflective dialogues, it is key to support innovators in self-defining their learning journeys, for example through sharing the collaborative learnings from the reflective process and translating them into actionable steps. However, not all insights are made explicit and might consequently, not be fully captured by simply noting down conclusions, which on the contrary may oversimplify rich information. It seems therefore important to not only record the main answers and conclusions but also gather the richness of information generated in the discussions by, for example, recording and transcribing dialogues and assigning a note-taker for the session. Next to understanding their learning needs, innovators should be able to plan actionable steps towards acquiring the new capabilities needed; our study showed that this is not a straightforward step for participants. A possible approach to facilitate this task is to guide participants to think about contacts in their network or context that possess such capabilities and could help in acquiring them. Identifying a personal learning network, as already pointed out in literature [9], may provide a path for the innovators to develop their learning journey after the reflection activity.

Even when recognising the value of reflection, it still seems challenging for innovation initiatives to introduce it systematically in their everyday practice. Insights from this study have shown the need to shorten a reflective session to fit within the time constraints of the design team workflow or offer a modular version of such activity that adapts to the timely needs and wants of the innovators. Despite the iterations that have been made for the current study, further research is needed to understand how to structure a reflective activity for it to be embraced as a systematic habit in innovators' practice. Another aspect worth considering is that the research done for this study happened mostly online. On the one hand, this permitted us to investigate how to structure reflections in a digital setting and inform possible ways to enable collaborative reflective activities from remote. On the other hand, further research could investigate how to introduce reflective activities in innovators' practice taking into consideration the everyday physical settings and contexts where that innovation practice is carried out.

2.6 Conclusions

The present study investigated how a reflective process could be structured to enable urban innovators to autonomously develop their learning trajectories and contribute to a better city. By employing reflective activities with several urban initiatives, we could identify building blocks for a reflective process that could help urban innovators frame new capabilities relevant to learn for the advancement of their practice. It can be concluded that the developed reflective processes showed valuable to urban innovators in unveiling new learning needs, but not necessarily in translating their learning into actionable steps in order to self-sustain their learning journey. Further research can be done to support urban innovators in learning from and with cities, to ultimately exploit them as learning ecosystems.

Acknowledgements The current study is part of the project DESIGNSCAPES (Building Capacity for Design-enabled Innovation in Urban Environments) funded by the Horizon2020 call CO-CREATION-02-2016—User-driven innovation: value creation through design-enabled innovation, under Grant Agreement No. 763784. The authors would like to thank the selected initiatives from the 2nd Open Call for Prototypes in the DESIGNSCAPES project for their participation in the study. Moreover, the support and fruitful discussions among the partners of the DESIGNSCAPES consortium are gratefully acknowledged.

References

1. Concilio, G., Tosoni, I.: Introduction. In: Concilio, G., Tosoni, I. (eds.) Innovation Capacity and the City. The Enabling Role of Design, SpringerBriefs in Applied Sciences and Technology, pp. 1–14. Springer, Cham (2019)
2. Concilio, G., Cullen J., Tosoni I.: Design Enabled Innovation in Urban Environments. In: Concilio, G., Tosoni, I. (eds.) Innovation Capacity and the City. The Enabling Role of Design, SpringerBriefs in Applied Sciences and Technology, pp. 85–101. Springer, Cham (2019)
3. Schon, D.A.: Educating the Reflective Practitioner. Toward a New Design for Teaching and Learning in the Professions. The Jossey-Bass Series in Higher Education. Jossey-Bass, San Francisco (1987)
4. Gray, D.E.: Facilitating management learning: Developing critical reflection through reflective tools. Manag. Learn. **38**(5), 495–517 (2007)
5. Helyer, R.: Learning through reflection: the critical role of reflection in work-based learning (WBL). J. Work-Appl. Manage. **7**(1), 15–27 (2015)
6. Dorst, K., Cross, N.: Creativity in the design process: co-evolution of problem–solution. Des. Stud. **22**(5), 425–437 (2001)
7. Magni, A.: Supporting Urban Innovators in Framing Their Capacity-Building Journey. MSc thesis Design for Interaction, Delft University of Technology (2020). Retrieved from: http://res olver.tudelft.nl/uuid:4f9f5bcc-f008-43bc-aadd-b3743d44d5ad
8. Gibbs, G.: Learning by Doing: A Guide to Teaching and Learning Methods. Oxford Further Education Unit, Oxford (1988)
9. Rajagopal, K., Joosten-ten Brinke, D., Van Bruggen, J., Sloep, P.B.: Understanding personal learning networks: their structure, content and the networking skills needed to optimally use them. First Monday (2012)

Chapter 3
Fighting the Gender Gap in ICT

Guidelines for Game Design

Cathrine Akre-Aas, Ingrid Kindem, and Monica Divitini

Abstract Girls and women are still under-represented in ICT, both in education and the tech industry. In this paper, we investigate how to design games for teenager girls to address the identified gender gap. Based on existing knowledge about this societal challenge, we identify four learning goals that games can address: promote self-confidence; fight gender stereotypes; boost subject knowledge; and provide role models. In addition, based on a systematic literature review, we identify game elements that are reported in previous research as having a positive impact on girls' game experience. The learning goals and game elements are summarized in a set of 16 design guidelines. The design guidelines are intended for game designers and developers. In addition, they might be used by educators to reflect on the games to introduce in their classes.

Keywords Game design · Gender gap in ICT

3.1 Introduction

The limited number of women in information and communication technology (ICT), both in education and work, is a well-known problem [1, 2]. The increased digitalization of work and society is making this gap even more problematic, and as a result, a growing number of initiatives are addressing this issue. Many of these initiatives are targeting teenagers with the awareness that the gender gap in ICT must be addressed during school years. These initiatives are often involving complex learning ecosystems. For example, public and private bodies are organizing programming camps for girls; university and industry are offering to bring role models into schools.

C. Akre-Aas · I. Kindem · M. Divitini (✉)
Department of Information and Computer Science, NTNU, Trondheim, Norway
e-mail: divitini@ntnu.no

C. Akre-Aas
e-mail: cathrbak@ntnu.no

I. Kindem
e-mail: ingriak@ntnu.no

Games have been proposed as one of the tools to help fighting the gender gap among teenager and as a promising approach to improve girls' attitudes regarding pursuing a career in computer science [3, 4]. However, it is important to remember that girls and boys have different preferences related to games [5]. For instance, [6] reports that girls prefer games with collaboration, while boys tend to prefer competition and individual play. These differences are important to take into account when designing games, but this body of knowledge is not easily available.

Contributing to this body of literature, the research question addressed in this paper is *how to design games that can help to address the gender gap in ICT among teenagers?* The paper contributes to the existing literature by providing a set of guidelines to support the design of games to fight the gender gap in ICT.

The paper starts by presenting four learning objectives that can be integrated in games to help fight the gender gap in ICT. Section 3.3 presents the results of a systematic literature review to identify relevant game elements. Section 3.4 brings together the learning objectives and game elements to create a set of guidelines. Section 3.5 presents a scenario of use for the guidelines, and Sect. 3.6 concludes the paper.

3.2   Learning Goals of the Game

Different factors influence the gender gap in ICT [7]. Based on existing literature on challenges to women involvement with ICT, we identify four main learning goals that games can address to help fighting the gender gap. These goals are closely connected, and they are distinguished mainly for analytical purposes. They are not intended to necessarily co-exist in the same game or to be addressed at the same level. They are mainly intended to create awareness that games can do more than helping girls to learn programming and help designers to adopt a broader perspective on the potentials of games.

LG1. Promote Self-Confidence. A critical issue in the early stage of the gender gap in ICT is the lack of confidence in one's own abilities. Some girls are undervaluing their achievements and think that they are unable to compete with boys in school because they believe that boys are better in STEM subjects than they are [8]. Girls will not pursue an ICT career if they believe that they would not succeed in it [9]. For example, studies of coding clubs show that lack of confidence might be one of the reasons causing low participation [10]. It is therefore important that a game to fight the gender gap is considering this issue and aims at improving girls´ self-confidence.

LG2. Fight Gender Stereotypes. To assume that men are more suited for scientific work, and that boys are better than girls in STEM, contributes to gender stereotypes, already in elementary schools [11, 12]. Gender stereotypes are highly connected to lack of confidence. When designing games, it is paramount to avoid confirming these stereotypes, but also design explicitly to increase in the player the awareness about the existence of these stereotypes and developing strategies to cope with them.

LG3. Boost Subject Knowledge. Girls' reluctance to engage with ICT might lead to a knowledge gap that is difficult to address while progressing with studies. For example, girls, who have not attended school or extra-curricular activities for learning programming early in their studies, might face a barrier when selecting more advanced courses. However, the problem does not only connect to lack of knowledge of the concepts and tools of ICT. Often, it connects more generally to the lack of understanding of what the field is and subject stereotypes. For example, in the interviews reported in [13], female secondary school students stated that they wished to help people in their future profession. Unfortunately, no one thought this was possible with ICT. Similar results are reported in a study by Microsoft [14]. In the same study, 9 out of 10 girls described themselves as creative. However, only 37% of the girls associated STEM professions with something that involves creativity. To summarize, when designing a game to fight the gender gap, promoting subject knowledge has to be carefully considered, addressing different perspectives. Promote subject knowledge is not only related to teaching programming but also to help the players to understand the field, including its social impact and creative nature.

LG4. Provide Role Models. The considerable gender gap in STEM and ICT, in particular, leads to the lack for many girls of role models that they can relate to. This is critical because role models can inspire girls and increase their interest in engaging with the subject [9, 15]. A game might help increase awareness about role models by providing, for example, information about or access to potential role models.

3.3 Game Elements for Girls

In this chapter, we present a review of the literature to identify the game elements that can help to design games that are suitable for girls.

3.3.1 Method Description

The goal of the literature review is to understand which game elements are known to promote the design of games that girls enjoy. The databases searched were IEEE Xplore, ACM digital library, Elsevier's Science Direct, Scopus, and ISI Web of Science. After some calibration, the final query was as follows: ("game design" OR "game elements" OR "serious games") AND (girl* OR teen* OR youth* OR female* OR gender*). The search resulted in 1707 articles (IEEE:114; ACM: 68, Elsevier's Science Direct: 48; Scopus: 1041; and ISI Web of Science: 436). The following eligibility criteria were defined for the screening of the papers:

- Report Eligibility—RE1: The publication must be in English; RE2: The publication must have an abstract; RE3: The publication must be published in 2010 or later; RE4: The publication should be published in international conferences, peer-reviewed journals, or as book chapters; RE5: The study must be accessible in full text without any fee.

- Study Eligibility—SE1: The publication must include a study of girls or gender differences among children; SE2: The publication must evaluate a game design or game elements or be a literature review on the topic.

To promote internal validity two of the authors screened a first set of 50 papers to make sure that the criteria are applied consistently. After checking the compatibility of the screening on this first set, the two authors proceeded separately dividing the remaining papers. The screening was first done on title and abstract, resulting in 99 papers. Both authors have then done the screening of these papers based on full text. This phase resulted in 24 relevant papers, 8 literature reviews, and 16 research papers.

The identified existing literature reviews are not analysed here in details. It is however worth to mention three of them that are relevant for the study and that are used in shaping our guidelines. Schwarz et al. [16] focus on the literature that is associated with user engagement in games for promoting healthy lifestyles among teenagers. Analysing 60 studies, the review identifies several game elements to create engagement among youths. Even though most game elements are described regardless of gender, some gender differences were pointed out [16]. Alserri et al. [17] analysed the existing literature on promoting higher female participation in Science, Technology, Engineering, and Mathematics (STEM), extracting 20 game elements matching with gender preferences in digital games. However, the target group for the games is not specified. Sharma et al. [4] present a literature review that investigates the relationship between game playing or design activities and girls' perception of computer science as a career choice. The systematic literature review's key conclusions are that ICT games for girls require personalization, opportunities for collaboration, and the presence of a female leads character. However, the authors point out that the strength of evidence is low as the time span of the interventions, and evaluation was limited. Therefore, no other specific advice to game and activity designers were offered.

A common issue in the existing literature reviews is that the studies they examine lack an explicit evaluation of the game elements. By taking into account the lessons learned from the identifies reviews, and narrowing down the search to fit our target group, we aim to contribute with knowledge on which game elements that are known to have an impact on teenagers girls.

3.3.2 Identifying Game Elements

The 16 papers identified after the screening of the search results were analysed using the mechanics-dynamics-aesthetics MDA framework [18]. The same framework is adopted for analysis also in [16], with the difference that their analysis does not focus on gender issues. (To ensure reliability in the process, the articles were coded independently by two of the two authors. After comparing the coding, all divergencies were discussed to reach an agreement.) The MDA framework can be used to analyse games by breaking the game elements into three categories. *Mechanics* describe the particular components of the game, at the level of data representation and algorithms. Together with the game content, it supports the overall gameplay dynamics. An example of a mechanic is a level. *Dynamics* describe the runtime behaviour, the interaction, between the users and the mechanics of the game. Besides, dynamics support the creation of aesthetic experiences. For instance, rewards can create a surprise experience, which is an aesthetic. The *aesthetics* describe the emotional response in the player when playing the game. Considering space limitations, we do not provide an overview of the single papers, but we report only the analysis following the MDA framework.

Mechanics. Six of the MDA framework mechanics were identified in the papers.

- **Avatar**. Avatar is the game mechanic discussed most frequently, including fantasy and animal characters [19, 20] as well as, in the majority of papers, more realistic ones [3, 21–24]. Furthermore, the importance of a player's resemblance and identification with the avatar was found in both [23, 25]. This is also supported by the findings on the dynamic component self-expression. We summarize with *M1:Realistic avatars in games, which the players can relate to or identify with*, and *M2:Sexualisation of female characters should be avoided*.
- **Hints**. [22, 26, 27] include hints in their game. Also, [24] recommends implementing mechanisms to strengthen self-efficacy in mastering technology tasks. Hints could be useful to prevent girls from feeling lost or defeated if they are stuck on a task. Even though there is not much evidence on hints nor on how they should be designed, we suggest to attempt **M3**: *Incorporate hints in the game*, since strengthening girls' self-efficacy is important to secure their enjoyment of the game.
- **Status and rewards**. Five papers [20, 22, 24, 27, 28] use levels in the game or evaluates this as a positive element. Further, three of these suggest that the difficulty should increase as one advances [22, 24, 27]. Based on these findings, we suggest **M4**: *Use levels in games and make the difficulty increasing throughout the levels*. However, there should be **M4a**: *A mechanic for switching between game levels* [28], or **M4b**: *The possibility of getting help when the player needs it* [27], so the challenge does not weaken girls' confidence. The most common mechanic to use as rewards were points. However, other mechanics like stars and positive encouragement were also found in [27]. Therefore, there is no clear pattern for how rewards should be fulfilled. Still, we argue that **M5**: *Rewards should be incorporated to encourage and reward the player*, to motivate and increase girls' confidence.
- **Devices and Game Control**. The analysed papers do not support any conclusion on these two mechanics.

Dynamics. After analysing the papers, seven different game dynamics were identified.

Guidance. Guidance was used in different ways in five papers. Dele-Ajayi et al. [22] provided the participants with a step-by-step guide booklet but found that the children preferred to experiment and try their own way of doing things. Textual guidance in the form of signposts or instructions attached was rare in the games created by the children in [26]. However, around half of the games contained some sort of instruction in conversations, paths or clues in the landscape. De Vette et al. [29] state that the games should be sufficiently challenging because it can enable achievement. At the same time, promoting self-efficacy is important to avoid players to get stuck [24]. Hints that players can access when they need it might be a useful element, as used in [24, 27]. To summarize, we suggest, **D1**: *Hints as the mechanism to fulfil the guidance dynamic.* By using hints, the players would not get step-by-step guidance, but still be able to access help if needed, to secure confidence.

Feedback. Feedbacks can be associated with something positive, like achievements [24, 29]. Robertson [26] found that girls were more interested in receiving feedback than boys, and Emembolu et al. [3] found that girls were more likely to take some action in response to comments than boys. This is possibly explaining why their final products were rated more highly. Hence, we conclude that both positive and constructive feedback are suitable in games for girls: **D2**: *Positive feedback to increase the motivation*, and **D3**: *Constructive feedback to increase the learning outcome* should be included.

Self-expression. Self-expression was identified in six of the papers and realized through the player's resemblance or identification with the main character. To be able to fulfil a player's identification with an avatar, there has to be a possibility of customizing the avatar or choosing among a variety of pre-made avatars. The advantage of the first solution is that the resemblance will be closer to the actual person, and the player will have the ability to make the character how she wants. However, this demands more resources in terms of development, so it is a choice that has to be considered. In the game used in [23], the female participants had difficulties of identifying with the game characters provided. This finding indicates that choosing among a variety of avatars might not be sufficient for securing self-expression in a game. Therefore, we will encourage to implement **D4**: *The possibility of customizing the avatars*, which includes both the main character and other important avatars, as the latter has been suggested in [25]. In addition to be able to choose how an avatar looks, [21] points out that one also should pay attention to a girl's ethnicity, interest, and motivation.

Status. Only few papers state explicitly something regarding the game's status; therefore, it is hard to conclude on an appropriate mechanic. However, several of the games used levels, which could be a suitable mechanic for the status. Therefore, we suggest using **D5**: *Levels for indicating the status of the game*, but we stress that this might not be the right element for all kinds of games.

Goal. Only four papers included a goal in their game. If a serious game is developed, the literature suggests that the learning objective should be the goal, as it is in [24, 25, 27]. Besides this, little can be said about the goal of the game from the literature provided. Thus, we summarize **D6**: *If serious game: learning objective as the goal of the game.*

Achievements and Rewards. As discussed, self-efficacy and confidence are important for girls. It can be supported with positive achievements and rewards. De Vette et al. [29] suggest to "provide content that enables achievement", like content that can be unlocked, to challenge the player continuously. Other games provide positive encouragement, points, stars, or positive feedback [20, 22, 27]. We suggest that **D7**: *Positive achievements and rewards* to be incorporated in the game.

Aesthetics. Seven categories were identified in the articles.

- **Sensation.** There were many suggestions in the articles, ranging from a total pink girl game [19], to other colour schemes. Thus, no clear pattern is found regarding sensation. However, it seems important that the graphics are designed wisely when designing for girls, as it is an important game feature for female players [20, 28]. Hence, the aesthetic **A1**: *The game should have high-quality graphics.*
- **Fantasy.** A realistic context seems to be popular among girls [6, 24, 25]. Similarly, Speiler and Slany [20] found that girls liked nature-themed graphics. However, the type of game world is tightly connected to the context of the game and should be designed accordingly. This leads to the aesthetic **A2**: *Realistic game world.*
- **Narrative.** It seems like violent content is disliked by female players, as it is mentioned in five of the articles in this review. Additionally, two articles propose that the narrative is an important game aesthetic for girls [20, 28]. This resulted in the aesthetic suggestion **A3**: *Non-violent content.*
- **Challenge.** Girls preferred different types of challenges. Some articles state that girls liked puzzles [20, 30], while others stated that puzzles were disliked most, and memory games were the most popular with girls [28]. Despite that, it seems evident that problem solving is a fun game element among girls. The aesthetic **A4**: *Include problem-solving tasks* was therefore created.
- **Fellowship.** Some of the papers suggest that a collaborative game approach is good for girls [6, 25, 27]. Additionally, Cunningham [21] stated that girls did not like competitive games. On the other hand, de Vette et al. [29] state that both competition and collaboration can be engaging for both genders, and that children, in general, like to play together. Social interaction is also proposed. A clear conclusion regarding fellowship cannot be stated. There are arguments for both collaborative and competitive games. However, it seems important to include social interaction in the game. A consequence of this is that the game can *A5: Include collaborative or competitive aspects,* but that it should *A6: Include social interaction.*
- **Discovery and Expression.** Though these game elements were mentioned, the analysed papers do not support any conclusion about them.

3.4 Design Guidelines

This section discusses how the implications summarized from the literature review can help to design games that meet the identified learning goals (Sect. 3.2), crystalizing this information into a set of guidelines.

Promote self-confidence. In order to gain confidence through the game, game mechanics such as hints (M3), levels (M4), and rewards (M5) can be included. Getting stuck in the game can decrease confidence and discourage the player. Providing guidance (D1) in the form of hints so that the player can move forward can prevent this from happening (M4b). Additionally, including positive rewards or achievements can create confidence. The dynamic game element feedback can also increase a player's confidence by providing positive feedback (D2) on the player's performance. Including some status (D5) in the game can also have a positive effect on the player's confidence, as it provides a clear overview and visualization of the progression in the game. However, if the player does not understand the game, the lack of positive feedback can create low confidence, so levels might be useful to prevent that. Using levels with increasing difficulty can also increase confidence, as the player can choose to play the level that corresponds to her abilities. The dynamic self-expression (D4) could also be used to increase a players confidence. By designing an avatar (M1) which one could identify with, the sense of achievement after mastering tasks can increase, since the player can see herself doing the tasks.

To conclude, the guidelines to promote self-confidence in ICT are as follows: **G1a**: Guidance through hints to proceed in the game; **G2**: Positive rewards or achievements to increase confidence; **G3**: Positive feedback on player's performance; **G4**: Status as a visualization of learning progression; and **G5a**: Customization of the player's avatar to identify with the avatar.

Fight Gender Stereotypes. To fight the stereotypes in ICT, including a realistic game world with realistic avatars (A2) with the presence of all types of programmers, from female programmers to the nerdy boys, can show the diversity that actually exists in the field. Further, graphics (A1) are identified as an important aesthetic for girls. The graphics can be utilized to showcase a diversity of people. It was found that self-expression was an important game dynamic, which could be fulfilled through the customisation of a player's avatar (D4). This could contribute to disproving the assumptions about stereotypes, by including an avatar which one resembles or identifies (M1) within the game.

In conclusion, the game elements that can help fighting stereotypes are as follows: **G6**: Realistic game world with realistic avatars that show the diversity in the field; **G7**: Good graphics which combat the stereotypes; and **G5b**: Customization of the player's avatar to remove stereotypical images of programmers.

Boost Subject Knowledge. Similarly to the game elements used to promote self-confidence, hints (M3), learning goal (D6), and constructive feedback (D3) can have a positive effect to boost knowledge. Firstly, it might be useful to consider games that facilitate learning programming, as recommended in [4, 17] to promote interest in STEM careers. The game can create awareness and possibly increase the player's knowledge of ICT through problem-solving tasks (A4), possibly focusing on the societal impact and relevance of ICT. As described in Sect. 3.2, most girls did not associate STEM jobs with creative tasks. Thus, even if the literature review has not identified creativity as an issue, it might be useful to integrate in the game creative tasks to rectify the impression that many girls have. Additionally, guidance (D1) in the form of hints can provide the necessary knowledge for the player to proceed with the tasks. The hints can, for instance, contain descriptions of unknown concepts to the player. Giving constructive feedback can also help the player to reflect, which is associated with increased knowledge. Lastly, including a status (D5) in the game can be a visualization on how much knowledge the player has gained.

The guidelines to boost the knowledge of the player are as follows: **G8**: Learning programming as focus of the game; **G9**: Problem-solving tasks to facilitate learning programming, with focus on the societal impact and relevance of ICT; **G10**: Incorporation of creative tasks; **G1b**: Guidance through hints for assistance in the game; and **G11**: Constructive feedback on tasks with the goal of triggering reflection; **G4**.

Provide Role Models. In the literature, we found that girls would benefit having role models. Role models could be female teachers, parents, or other female role models in STEM. Both positive (D2) and constructive (D3) feedback were found to be an effective game dynamic. Further, Spangenberger et al. [24] found that non-player characters play a crucial role as girls' feedback providers. Therefore, we suggest using role models as feedback providers. Also, girls want to make their own choices regarding characters. Thus, a spectre of role models to choose between (D4) would be suitable, so the player could choose the one she identifies with (M1) or admires the most as her feedback provider. Having a role model as the feedback provider could also be used to fight stereotypes. To conclude, the guidelines for providing role models are as follows: **G12**: Non-player-character: Design a range of avatars that the player can choose to be the Non-player characters (NPC); and **G13**: Feedback provider: Use the selected NPC. Design so that the NPC can act as a role model and motivator.

In addition, there are three guidelines that are more general. To prevent isolation, playing a collaborative (**A5**) game could create a feeling of being included. By letting the girls play with other girls, they can get a feeling of belonging in the ICT field together through social interaction (**A6**) with each other. Providing social interaction can also fight stereotypes, as some think that being a programmer is an asocial profession. Thus, the guideline to facilitate social interaction will be the following: **G14**: Collaborative gameplay. We found that non-violent content (**A3**), in addition to the sexualisation of female characters (**M2**), should be avoided as there were strong indications against it. Therefore, two more guidelines will be applied, which are as follows: **G15**: Non-violent content; and **G16**: Avoid sexualisation of female characters.

3.5   Using the Guidelines to Compare Programming Games

Design guidelines are primarily intended as a tool for game designers and developers in supporting the creation of games. However, the guidelines could also be used by educators to compare and reflect on programming games and support them in selecting games to be used in their activities. Let us illustrate this use with two examples. Consider two rather popular programming games: CodeCombat (https://codecombat.com) and CodeMonkey (https://codemonkey.com).

CodeCombat is based on teaching text-based-programming through a number of tasks with increasing difficulty in a game world. CodeCombat is specifically designed for use in a class context, where the teacher can get an overview of the progress of each student. CodeCombat fulfils approximately six of the 16 guidelines. The game has hints, rewards, achievements, status, and thus fulfilling G1, G2, and G3. The player can change avatar, but not promoting full identification. In particular, the available avatars will not remove stereotypical images of programmers, they will more likely strengthen them (G5b). However, the game has problem-solving tasks to facilitate learning programming G9, as well as providing constructive feedback to trigger reflection as in G11. So, though the game might be a useful tool to learn programming, there are some concerns about its design when it comes to promote girls´ participation. This is in line with existing research reporting that girls did not feel welcome in the masculine game environment in CodeCombat [23].

CodeMonkey (https://codemonkey.com) is a platform offering different games. Among others, it contains games which have text-based, block-based, and Python-based courses. Because of the different difficulty levels, no previous coding experience is needed to start playing on the platform. Courses are designed for school, extra-curriculum, or home use. In addition to offer a learning platform for individuals, a class context is offered, where teachers have the possibility of following the students' progress. The game fulfils multiple of the guidelines, to sum up: G1, G2, G3, G4, G9, G10, G15, and G16. Before playing the game, an avatar has to be chosen. The different avatars range from animals to humans, in different shapes and looks. Therefore, one could identify with the avatar (G5a) and remove the stereotypical images of programmers (G5b). However, this avatar is almost invisible in the game. Only a small picture of it is visible in the top right corner, where you can click it to access the profile. Therefore, since the avatar the player chose is not the main player-character, G5 cannot be seen as achieved. Since the game unfolds in a jungle environment, the realistic game world described in G6 is not achieved. Moreover, the graphics are good, but do not combat stereotypes; hence, G7 is not followed. The game revolves around teaching programming, not about awareness of ICT; therefore, G8 is not fulfilled. Two students could collaborate on solving the tasks, but there does not exist a built-in function that facilitates for students to play online together; hence, G14 is not followed.

This assessment is not intended to exclude any game from the set of tools adopted by educators. However, an understanding of the flaws in the design with respect to gender might increase awareness of teachers and push them to introduce corrective actions to compensate for the limitations. It should also be noted that most of the commercial games are focusing on learning programming. Educators should consider integrating with other games or activities that address more directly the other learning goals identified in Sect. 3.2. For example, the game described in [31] does not aim at teaching programming, but rather focuses on a game concept aiming at explaining the role that technology plays in everyday life, fighting stereotypes connected to ICT.

3.6 Conclusions

The paper contributes to the body of literature on games for addressing the gender gap in ICT. Based on existing knowledge about this societal challenge, we identify four learning goals that games can address; promote self-confidence; fight gender stereotypes; boost subject knowledge; and provide role models. In addition, based on a systematic literature review, we identify game elements that are reported in previous research as having a positive impact on girls' game experience. The learning goals and game elements are summarized in a set of 16 design guidelines.

The guidelines are not prescriptive and are mainly intended as a way to summarize existing knowledge in the field. Future research is aiming at identifying how to integrate the various guidelines in games in an effective way. Is there any combination of learning objectives and game elements that is particularly effective or challenging? Also, most of the existing research is not about games in the area of ICT. So, future research is needed to refine the guidelines for the specific area of concern.

Acknowledgements The work is partly funded by Excited, The Norwegian Centre for Excellent IT Education (https://www.ntnu.edu/excited).

References

1. Ensmenger, N.: The Computer Boys Take Over: Computers, Programmers, and the Politics of Technical Expertise. The MIT Press (2010)
2. Beede, D.N., Julian, T.A., Langdon, D., McKittrick, G., Khan, B., Doms, M.E.: Women in STEM: A Gender Gap to Innovation. Social Science Research Network, Rochester, NY (2011). https://doi.org/10.2139/ssrn.1964782
3. Emembolu, I., Strachan, R., Davenport, C., Dele-Ajayi, O., Shimwell, J.: Encouraging Diversity in Computer Science Among Young People: Using a Games Design Intervention based on an Integrated Pedagogical Framework. Proceedings—Frontiers in Education Conference, FIE. 2019-Octob (2019). https://doi.org/10.1109/FIE43999.2019.9028436
4. Sharma, K., Torrado, J.C., Gómez, J., Jaccheri, L.: Improving Girls' Perception of Computer Science as a Viable Career Option Through Game Playing and Design: Lessons from a Systematic Literature Review. Elsevier B.V. (2021). https://doi.org/10.1016/j.entcom.2020.100387

5. Kafai, Y.B.: Considering gender in digital games: Implications for serious game designs in the learning sciences. In: Computer-Supported Collaborative Learning Conference, pp. 422–429. CSCL (2008)
6. Admiraal, W., Huizenga, J., Heemskerk, I., Kuiper, E., Volman, M., Dam, G.T.: Gender-inclusive game-based learning in secondary education. Int. J. Incl. Educ. **18**, 1208–1218 (2014). https://doi.org/10.1080/13603116.2014.885592
7. Cheryan, S., Ziegler, S.A., Montoya, A.K., Jiang, L.: Why are some STEM fields more gender balanced than others? Psychol. Bull. **143**, 1–35 (2017). https://doi.org/10.1037/bul0000052
8. Selimbegovic, L., Chatard, A., Mugny, G.: Can we encourage girls' mobility towards science-related careers? Disconfirming stereotype belief through expert influence. Eur. J. Psychol. Educ. **22** (2007). https://doi.org/10.1007/bf03173426
9. Hill, C., Corbett, C., Andresse, S.R.: Why So Few? (2010)
10. Corneliussen, H.G., Prøitz, L.: Kids Code in a rural village in Norway: could code clubs be a new arena for increasing girls' digital interest and competence? Inf. Commun. Soc. **19**, 95–110 (2016). https://doi.org/10.1080/1369118X.2015.1093529
11. Ambady, N., Shih, M., Kim, A., Pittinsky, T.: Stereotype susceptibility in children: effects of identity activation on quantitative performance. Psychol. Sci. **12**, 385–390 (2001). https://doi.org/10.1111/1467-9280.00371
12. Farenga, S.J., Joyce, B.A.: Intentions of young students to enroll in science courses in the future: an examination of gender differences. Sci. Educ. **83**(1), 55–75 (1999). https://doi.org/10.1002/(sici)1098-237x(199901)83:1%3c55::aid-sce3%3e3.0.co;2-o
13. Serussi, S., Divitini, M.: Girls and computing in lower secondary education the surprisingly unsurprising results of a Norwegian exploratory study. UDIT 2017 (2017)
14. Microsoft: Why Europe's girls aren't studying STEM (2018)
15. Black, J., Curzon, P., Myketiak, C., McOwan, P.W.: A study in engaging female students in computer science using role models. In: Proceedings of the 16th Annual Joint Conference on Innovation and Technology in Computer Science Education, pp. 63–67. Association for Computing Machinery, New York, NY, USA (2011). https://doi.org/10.1145/1999747.1999768
16. Schwarz, A.F., Huertas-Delgado, F.J., Cardon, G., Desmet, A.: Design features associated with user engagement in digital games for healthy lifestyle promotion in youth: a systematic review of qualitative and quantitative studies. Mary Ann Liebert Inc. (2020). https://doi.org/10.1089/g4h.2019.0058
17. Alserri, S.A., Zin, N.A.M., Wook, T.S.M.T.: Gender-based engagement model for serious games. Int. J. Adv. Sci. Eng. Inf. Technol. **8**, 1350–1357 (2018). https://doi.org/10.18517/ijaseit.8.4.6490
18. Hunicke, R., Leblanc, M., Zubek, R.: MDA: a formal approach to game design and game research. In: AAAI Workshop—Technical Report. WS-04-04, 1–5 (2004)
19. Khaled, R.: Equality = inequality: probing equality-centric design and development methodologies. Lecture Notes in Computer Science (including subseries Lecture Notes in Artificial Intelligence and Lecture Notes in Bioinformatics). 6947 LNCS, pp. 405–421 (2011). https://doi.org/10.1007/978-3-642-23771-3_30
20. Speiler, B., Slany, W.: Game Development-Based Learning Experience: Gender Differences in Game Design (2018)
21. Cunningham, C.: Girl game designers. New Media Soc. **13**, 1373–1388 (2011). https://doi.org/10.1177/1461444811410397
22. Dele-Ajayi, O., Shimwell, J., Emembolu, I., Strachan, R., Peers, M.: Exploring digital careers, stereotypes and diversity with young people through game design and implementation. In: IEEE Global Engineering Education Conference, EDUCON. 2018-April, pp. 712–719 (2019). https://doi.org/10.1109/EDUCON.2018.8363301
23. Yücel, Y., Rızvanoğlu, K.: Battling gender stereotypes: a user study of a code-learning game, "Code Combat", with middle school children. Comput. Hum. Behav. **99**, 352–365 (2019). https://doi.org/10.1016/j.chb.2019.05.029
24. Spangenberger, P., Kruse, L., Kapp, F.: Serious games as innovative approach to address gender differences in career choice. In: Lecture Notes in Computer Science. pp. 431–435. Springer (2019b). https://doi.org/10.1007/978-3-030-11548-7_43

25. Ochsner, A.: Lessons learned with girls, games, and design. In: ACM International Conference Proceeding Series. pp. 24–31. Association for Computing Machinery (2015)
26. Robertson, J.: Making games in the classroom: benefits and gender concerns. Comput. Educ. **59**, 385–398 (2012). https://doi.org/10.1016/j.compedu.2011.12.020
27. Hussein, M.H., Ow, S.H., Cheong, L.S., Thong, M.K.: A Digital game-based learning method to improve students' critical thinking skills in elementary science. IEEE Access. **7**, 96309–96318 (2019). https://doi.org/10.1109/ACCESS.2019.2929089
28. Tuparova, D., Tuparov, G., Veleva, V.: Girls' and Boys' Viewpoint on Educational Computer Games (2019). https://doi.org/10.34190/GBL.19.130
29. de Vette, A.F.A., Tabak, M., Vollenbroek-Hutten, M.M.R.: How to design game-based health-care applications for children? A study on children's game preferences. In: HEALTHINF 2018, pp. 422–430. SciTePress (2018)
30. Tatli, Z.: Traditional and digital game preferences of children: a CHAID analysis on middle school students. Contemp. Educ. Technol. **9**, 90–110 (2018)
31. Saxegaard, E., Divitini, M.: CITY: A game to raise girls' awareness about information technology. In: Pozdniakov, S.N., Dagienė, V. (eds.) Informatics in Schools. New Ideas in School Informatics. pp. 268–280. Springer International Publishing, Cham (2019)

Chapter 4
Toward the Set of Principles for Enhancing Open Cultural Data Reusability in Educational Context

Oleksandr Cherednychenko and Kai Pata

Abstract Open cultural data is comprised of digitized heritage artifacts, released by heritage institutions into open access. It constitutes a powerful tool to transform current educational practices, with applications ranging from textbook augmentation to digital games and data-based services. That transformation is achieved via data reuse, allowing reconceptualization of available datasets in creative ways. However, successful implementation of data reuse initiatives is possible only if technical prerequisites are met. Review of existing open data governance principles has revealed that aspects crucial for reuse are not covered, therefore hindering the wide adoption of reuse practices. In this paper, we present a set of principles to act as guidelines for enhancing the reusability potential of open cultural data. Proposed set of principles was utilized to analyze Estonian open cultural data landscape and formulate improvement suggestions. Our analysis has shown substantial gaps which should be resolved before wide-scale educational reuse of Estonian open cultural data that could be initiated.

Keywords Open cultural data · Data governance principles · Data reuse

4.1 Introduction

Starting from 1990s, the European Commission has supported multiple projects aimed at enabling access to heritage materials to stimulate a disruptive information society [1]. Cultural heritage could be perceived as "a source of social innovation for smart, sustainable and inclusive growth," and it is stated that "digitization and online accessibility of cultural content shake up traditional models, transform value chains and call for new approaches to our cultural and artistic heritage" [2]. According to the Declaration of Cooperation on Advancing Digitization of Cultural Heritage

O. Cherednychenko (✉) · K. Pata
School of Digital Technologies, Tallinn University, Narva mnt 25, 10120 Tallinn, Estonia
e-mail: oleksandr.cherednychenko@tlu.ee

K. Pata
e-mail: kai.pata@tlu.ee

© The Author(s), under exclusive license to Springer Nature Singapore Pte Ltd. 2022
Ó. Mealha et al. (eds.), *Ludic, Co-design and Tools Supporting Smart Learning Ecosystems and Smart Education*, Smart Innovation, Systems and Technologies 249,
https://doi.org/10.1007/978-981-16-3930-2_4

[3], "additional focus in overall digitization efforts should be placed on the reuse of digitized cultural resources," with the overall aim being "to foster citizen engagement, innovative use and spill-overs in other sectors." Heritage re-utilization is also one of the five action areas in the European Framework for Action on Cultural Heritage [4], described as "cultural heritage for an inclusive Europe: participation and access for all.".

The technological solutions developed for allowing access to digitized heritage are quite diverse and are represented by national aggregator portals and the pan-European platform titled Europeana. As for the national portals, they parse the data from diverse national data sources, generally being represented by GLAMs (galleries, libraries, archives and museums) and provide a common search-like interface to all the variety of national digitized cultural heritage. One example of national aggregator portal would be Finna.fi, which could be described as Finland's "service combining the materials of archives, libraries and museums" [5]. Europeana, in its turn, also operates as a metadata aggregator, combining the cultural data and serving as a single access point to digitized content from over 3200 European institutions [6]. It should be also noted that as highlighted by Concordia et al. [7], Europeana is intended to serve as open-access platform, exposing the rich application programming interface (API) to enable interoperable data flows between Europeana and content providers and opening the road for end-user enrichment of cultural artifacts.

This two-way data flow also closely resonates with the Europeana Strategy 2015–2020 [8], where it is outlined that "people want to reuse and play with the material, to interact with others and participate in creating something new." Therefore, the goal for Europeana is to start "behaving like a platform—a place not only to visit but also to build on, play in and create with" [8].

However, the utilization of digitized heritage became stalled by several obstacles, thus effectively preventing the non-expert users from extensively interacting with digital artifacts. This paper intends to introduce a set of principles that would enable wider open cultural data reuse with the educational domain.

4.2 Related Work

4.2.1 Open Cultural Data

UNESCO's Charter on the Preservation of Digital Heritage Charter on the Preservation of Digital Heritage [9] states that "the purpose of preserving the digital heritage is to ensure that it remains accessible to the public. Accordingly, access to digital heritage materials, especially those in the public domain, should be free of unreasonable restrictions." However, as it was noted by Borgman [10], "so what use are the digital libraries, if all they do is put digitally unusable information to the web?" The accessibility issues were researched by multiple authors and include overly restrictive data usage licenses [11], unfriendly user interfaces [12] and lack of heritage content specifically tailored for being reused by general public [12].

As it is specified in the Open Data Book, "Open data is data that can be freely used, reused and redistributed by anyone—subject only, at most, to the requirement to attribute and share alike." While this definition is generally applied to open government data, it could be argued that digitized heritage data is conceptually no different (when there is no copyright involved) and thus should also be made publicly assessable under the same conditions. Indeed, there also exists a dedicated OpenGLAM initiative OpenGLAM Foundation [4], which advocates for the opening of digitized artifacts. Highlighting the overall idea behind the OpenGLAM, it could be described as "allowing free reuse, digitized collections shift from passive showcases to become raw material for every user to enjoy, learn from and build upon" [13]. Needless to say that the further opening of museum collections would also mean the transition away from the traditional approach of institutions providing their digitized exhibits in curated collections to the wider variety of artifacts being available for the exploration.

Additionally, it should also be mentioned another issue with licensing is the confusion it causes—as it is not quite clear how could the objects be used and reused. Following the example by Terras [12], it should be said that it is of paramount importance that the heritage artifacts would have the license details specified—weather or not the particular object has been moved to the public domain and is thus exempt from all the licensing restrictions.

4.2.2 Cultural Data Reuse

As outlined in Europeana Strategy 2015–2020, "people want to reuse and play with the material, to interact with others and participate in creating something new" [8]. Based on the work by Terras [12], cultural data reuse approaches could be divided into two distinct categories, namely physical reuse and digital reuse.

Starting from the physical reuse, the overall idea is to move the digital objects into physical modality by using the artifacts for inspiration and materials for handicrafts, including, for example, phone cases or creative jewelry. One example of such reuse would be the USEUM project, which is an online crowdsourced art museum with a build-in e-commerce functionality—enabling, for example, to purchase a print of favorite painting [4].

However, the overall issue with the facilitation of physical reuse could be described as the lack of a suitable interface metaphor. Walsh and Hall [14] notice that the existing cultural heritage platforms mostly provide a search-box interface as a primary tool for accessing the cultural data. However, this type of interface is suitable mostly for experts looking for some specific artifact, but it not so welcoming for the casual user who is just browsing around. One solution, proposed by the authors, would be to switch the focus from supporting the search process to supporting the understanding of what is available in the collection and where the browsing might be started from Walsh and Hall [14]. Additional perspective on browsing style is introduced in the paper by Petras et al. [6], where it is stated that users belonging to the creative industries might also benefit from the unexpected search results being displayed, as their primary motivation for using the digital heritage platform is essentially quest for inspiration.

Furthermore, another problem specified by Terras [12] is the resolution of digitized artifacts, especially for the cases of pictorial heritage. Indeed, it is quite hard to print or otherwise utilize the low-resolution image, and it would be beneficial for the public if the heritage platforms would also allow to download the high-resolution pictures. Finally, it should be also noted that the problem of low-resolution artifacts is applicable to both physical and digital reuse.

Moving on to the digital reuse, it is mostly oriented toward the utilization of artifacts in third-party services, focused on applications in education [8] or tourism sectors [15]. From the technical perspective, the major solution supporting the digital reuse and interaction with third-party services is applied programming interface (API). However, while the availability of API enables the interaction between cultural heritage platform and external services, it should be also mentioned that the previously discussed lack of clarity in licensing should be sorted out on the platform level, and it would be reasonable to assume that only artifacts in public domain should be opened for reuse. While the heritage platforms currently suffer from the interface choices that are not especially friendly to the causal users (not heritage professionals), this limitation could potentially be solved by introducing the novel third-party services which would consume the cultural data from the platforms(s) and enable different modes of interaction.

4.2.3   Cultural Data Reuse in Education

Open cultural data has extensive potential for transforming the educational field. Truyen et al [15] describe Europeana, serving as primary storage of European open cultural data as being an "ideal tool," allowing public reuse to digitized collections providing diverse (as opposed to many current resources) materials that could be integrated into a variety of outputs, including MOOCs and educational apps. An overview of reusage projects, provided by Ferrara et al. [16], includes also lessons with integrated museum objects, interactive games and personalized video tours. Finally, another major domain of educational reuse is comprised of crowdsourcing initiatives, categorized by Oomen and Aroyo [17] into six subtypes with most relevant being correction and transcription, contextualization and complementing collections. It should be added that crowdsourcing in GLAM institutions is predominantly introduced via gamification [18]. While a significant number of gamified crowdsourcing initiatives exist, one of the primary challenges from the technical perspective is the inability to reinject back the enriched data, thus limiting the visibility and significance of crowdsourced aspects. Overall, it could be concluded that educational reuse takes all shapes and sizes, ranging from integrating artifacts into lesson materials to building dedicated educational apps. Common for all of them, however, is the need for technical accessibility and discoverability of open cultural data.

4.2.4   Data Governance Principles

Performed topical literature review has identified several widely adopted sets of principles, potentially applicable for governing the open cultural data. Starting from the top level of abstraction, cultural data fits under the umbrella term of open data, defined by Open Knowledge Foundation as data that "could be freely used, modified, and shared by anyone for any purpose" [14]. Compliant with the Open Knowledge Foundation's definition of openness, cultural data in particular is in focus of the OpenGLAM movement, which produced a set of principles specifically tailored to the cultural heritage domain [4]. Finally, another set of data management related principles is FAIR—introduced by Wilkinson et al. [19] to enhance the reusability of scientific data. Internal interdependencies between principles are not explicitly stated, but it could be argued that both OpenGLAM and FAIR principles should comply with Open Definition while introducing specifics relevant for their targeted scopes— cultural and scientific data, respectively. The details of aforementioned principles are summarized in Table 4.1.

4.3   Proposed Principles

As previously discussed, several sets of principles currently exist to establish the discoverability and reusability of data. However, while all of them explicitly cover the machine readability aspect—ensured via the adoption of metadata and open data formats, little attention is given to the human discoverability—or the possibility of end users to find and use the data. Drawing from an adjacent field of open government data, there exists a defined notion of data portal which makes the data available publicly and therefore serves as a primary interaction point between the data provider and the data user [20]. While several notable portal examples exist in the open cultural data domain, there seems to be no consensus within the community reflecting the importance of portals for unlocking the heritage to the wider population. Europeana, which could be perceived as one of the premier open cultural data portals, was initially conceived for enabling access to digital surrogates of heritage objects via a dedicated API, with Europeana Portal being a reference implementation of the API [7]. Since Europeana is not intending to become a world-wide open cultural data aggregator, it seems important to acknowledge the significance of data portals as tools enabling wider dissemination of cultural heritage data for relevant societal applications.

One more aspect, currently missing from the existing principles is the need for enabling data access via an API—specifically, read–write API as it's an "excellent strategy for the enrollment in or participation in government services" [8]. While the presented quote refers to open government data and government services in particular, it could be argued that specific type of services is not relevant from the implementation perspective, and the approach is also applicable to the open cultural data.

Therefore, in the attempt to resolve the specified issues, three existing sets of data governance principles (Open Definition, OpenGLAM and FAIR) were categorized and examined. As a result of the examination, we propose a set of reusability principles, as shown next (Table 4.2).

Table 4.1 Overview of data governance principles

Name	Scope	Principles
Open Definition [14]	Open data	Open license or status The work must be in the public domain or provided under an open license Access The work must be provided as a whole and at no more than a reasonable one-time reproduction cost and should be downloadable via the Internet without charge Machine readability The work must be provided in a form readily processable by a computer and where the individual elements of the work can be easily accessed and modified Open format The work must be provided in an open format—the one which places no restrictions, monetary or otherwise, upon its use and can be fully processed with at least one free/libre/open-source software tool
OpenGLAM [4]	Cultural heritage data	Release digital information about the artifacts (metadata) into the public domain using an appropriate legal tool Keep digital representations of works for which copyright has expired (public domain) in the public domain by not adding new rights to them When publishing data make an explicit and robust statement of your wishes and expectations with respect to reuse and repurposing of the descriptions, the whole data collection, and subsets of the collection When publishing data use open file formats which are machine-readable Opportunities to engage audiences in novel ways on the web should be pursued

(continued)

Table 4.1 (continued)

Name	Scope	Principles
FAIR [19]	Scientific data	Findable F1. (meta)data are assigned a globally unique and persistent identifier F2. data are described with rich metadata F3. metadata clearly and explicitly include the identifier of the data it describes F4. (meta)data are registered or indexed in a searchable resource Accessible A1. (meta)data are retrievable by their identifier using a standardized communications protocol A1.1. the protocol is open, free, and universally implementable A1.2. the protocol allows for an authentication and authorization procedure, where necessary A2. metadata are accessible, even when the data are no longer available Interoperable I1. (meta)data use a formal, accessible, shared and broadly applicable language for knowledge representation I2. (meta)data use vocabularies that follow FAIR principles I3. (meta)data include qualified references to other (meta)data Reusable R1. meta(data) are richly described with a plurality of accurate and relevant attributes R1.1. (meta)data are released with a clear and accessible data usage license R1.2. (meta)data are associated with detailed provenance R1.3. (meta)data meet domain-relevant community standards

4.4 Estonian Digital Heritage Landscape

With the proposed set of principles being in place, in this chapter, we attempt to utilize it for assessing the reuse potential of open cultural data in Estonia. Starting from the landscape overview, open cultural data is stored in multiple repositories based on the artefact origin. Since artifacts are predominantly managed by three memory institutions—National Library[1] (Eesti Rahvusraamatukogu), National Archives[2] (Rahvusarhiiv) and a variety of museums, open cultural data is stored in the respective institutional repositories. It should also be noted that museums are

---

[1] https://www.nlib.ee/.

[2] https://www.ra.ee/.

Table 4.2 Proposed principles

Use case	Tools	Principles
Access	Data portal	Data should be uploaded on the publicly available open cultural data portal
Find	Metadata, data model	Data should be comprehensively described by metadata, following one of the established heritage data models Metadata should be of sufficient completeness and quality
Retrieve	Open format, machine readability, API's	Data should be freely (without mandatory registration) available for download in one of open data formats Data should be available via a designed two-way (read-write) API
Utilize	Open license, explicit declaration of rights, public access	Data should be released under open license, explicitly outlining the usage conditions Data in public domain should have a respective identifier

connected to the central information system called MuIS,[3] enabling to access all the digitized museum collections across the country via a single technological solution [21]. Finally, there is also a universal E-repository[4] (E-Varamu) which was created as a single access point to all resources of memory institutions. The complete list of relevant repositories is available in Table 4.3.

While from the first glance, it looks like Estonian open cultural data which could be easily accessed and subsequently reused, in that is not the case. E-varamu, intended to act as an aggregator, stores only artefact metadata (similar to Europeana)—thereby prompting the user to access digitized artefact on the respective repository webpage. While not an issue by itself, certain repositories (namely DIGAR digital archive and SAAGA) require user identification, hence limiting the openness of data and its potential reusability. Current state of Estonian data aggregation in Europeana also does not allow to mitigate issues caused by suboptimal implementation of open cultural data infrastructure. According to Europeana Foundation (2021), 959,628 objects in the Europeana Collections are provided by the Estonian institutions. However, while those objects come from 25 institutions, almost half of those (413,072) are from Estonian War Museum—resulting in a data skew.

Therefore, it could be concluded that Estonian cultural data is currently lagging in the realm of retrievability—as certain data repositories require user registration to access digitized artifacts. Even with all the other functionalities being in place, lack of truly open-access prevents published cultural data from being reused. Moreover, another major issue is the complete absence of data API's—resulting in service

---

[3] https://www.muis.ee/.

[4] https://www.e-varamu.ee/.

Table 4.3 Estonian open cultural data repositories

Repository	Scope	Institution	Access	Retrieval	Website
DIGAR digital archive	E-books, newspapers (before 1821), magazines (before 2017), maps, sheet music, photos, postcards, posters, illustrations, audiobooks, and music files	National Library	Partially restricted, user identification for specific artifacts	Depending on artifacts, single object download, no public API	https://www.digar.ee/arhiiv
DIGAR Estonian articles	Magazines (after 2017) and newspapers (after 1821)	National Library	Open	Single object download, no public API	https://dea.digar.ee/cgi-bin/dea
MuIS	Museum collections from 60 museums (2021)	National Heritage Board	Open	Single object download, no public API	https://www.muis.ee/
SAAGA	Digitized archival sources	National Archives	Restricted, user identification	Single object download possible after registration, no public API	https://www.ra.ee/dgs/explorer.php
FOTIS	Photographs	National Archives	Open		https://www.ra.ee/fotis/
KARDID	Digitized maps	National Archives	Open	Single object download, no public API	https://www.ra.ee/kaardid/
E-varamu	Shared—artifacts from Estonian libraries, museums and archives	Independent	Open	Possible via redirect to respective repository	https://www.e-varamu.ee/

development being impossible. This also results in dubious crowdsourcing practices—while several crowdsourcing initiatives (Municipal Courts[5] and War of Independence[6]) were pioneered by National Archives, there is no user-level technical opportunity to re-ingest enriched data. Therefore, in its current from Estonian repositories do not support reuse initiatives, meaning that several changes must be made. One specific proposal would be to further extend national aggregator (E-varamu)—via adding API functionality and enabling it to truly serve as single point of contact for all cultural heritage-related inquiries. That approach has already been proven in

[5] https://www.ra.ee/vallakohtud/

[6] https://www.ra.ee/vabadussoda.

Table 4.4 Proposed changes

Use case	Improvements
Access	Existing data portal should be comprehensive
Find	Metadata should have minimal completeness requirements—basic set of descriptors is needed for any artefact
Retrieve	Registration-only access to data repositories should be abolished Data APIs should be introduced to enable data services development
Utilize	License should be a mandatory metadata field and should be specified for any digitized artefact

practice, for example, in Finland's Finna.fi [5]. From the overall policy perspective, it could be noted that there should be a conceptual turn from focusing on merely digitizing all national artifacts to ensuring that digital artifacts could be easily accessed and further reused. Finally, summary of proposed high-level changes is presented in Table 4.4.

4.5 Conclusions and Further Work

This paper explores the challenges currently preventing open cultural data from being extensively reused for educational purposes by providing a set of principles which should technically enable the reuse functionalities. While proposed set of principles presumably aids in reusability and was evaluated on Estonian cultural data landscape, further research is needed to fully assess its applicability in practical setting. It should be also noted that assessment of Estonian cultural data landscape has revealed several deficiencies which could be mitigated by modifying existing technological solution. It could be assumed that performing analysis on the solution design stage would simplify the changes and reduce their cost.

Several differentiating principles, namely criticality of data portals and read–write APIs, were inferred from open government data research and not widely discussed within the open cultural data body of knowledge. Concept of open data encompasses both government and cultural data, rendering them indistinguishable from the technical perspective. Yet, additional research is needed to establish whether those notions are indeed befitting for the GLAM sector.

Finally, this study covered reusability from a technical perspective, leaving out other factors and obstacles that could be emerged during large scale adoption of cultural data reuse. Therefore, compliance with data governance principles is a prerequisite but definitely not the only challenge.

References

1. Borowiecki, K.J., Navarrete, T.: Digitization of heritage collections as indicator of innovation. Econ. Innov. New Technol. **26**(3), 227–246 (2017)
2. Whitehand, J.: Towards an integrated approach to cultural heritage for Europe. In: Communication from the Commission to the European Parliament, the Council, the European Economic and Social Committee and the Committee of the Regions, COM, p. 477 (2014)
3. European Comission: Declaration of Cooperation on Advancing Digitisation of Cultural Heritage (2019). https://www.vi-mm.eu/wp-content/uploads/2016/12/scannedsignedDeclaration090419pdf.pdf
4. Valeonti, F.: USEUM: Making Art Accessible with Crowdsourcing and Gamification (2018)
5. Tolonen, E.: Finna in a Nutshell (2017)
6. Petras, V., Hill, T., Stiller, J., Gäde, M.: Europeana—a search engine for digitised cultural heritage material. Datenbank-Spektrum **17**(1), 41–46 (2017)
7. Concordia, C., Gradmann, S., Siebinga, S.: Not just another portal, not just another digital library: a portrait of Europeana as an application program interface. IFLA J. **36**(1), 61–69 (2010)
8. Tauberer, J.: Open Goverment Data: The Book (2014)
9. UNESCO: Charter on the Preservation of Digital Heritage (2003). http://portal.unesco.org/en/ev.php-URL_ID=17721&URL_DO=DO_TOPIC&URL_SECTION=201.html
10. Borgman, C.L.: The Digital Future is Now: A Call to Action for the Humanities (2010)
11. Eschenfelder, K.R., Caswell, M.: Digital cultural collections in an age of reuse and remixes. Proc. Am. Soc. Inf. Sci. Technol. **47**(1), 1–10 (2010)
12. Terras, M.: So you want to reuse digital heritage content in a creative context? Good luck with that. Art Libr. J. **40**(4), 33–37 (2015)
13. Glasemann, K.: Inside the Museum is Outside the Museum—Thoughts on Open Access and Organisational Culture. Open GLAM (2020, March 13). https://medium.com/open-glam/inside-the-museum-is-outside-the-museum-thoughts-on-open-access-and-organisational-culture-1e9780d6385b
14. Walsh, D., Hall, M.: Just Looking Around: Supporting Casual Users Initial Encounters with Digital Cultural Heritage, p. 1338 (2015)
15. Truyen, F., Colangelo, C., Taes, S.: What can Europeana bring to open education? Enhancing European Higher Education "Opportunities and Impact of New Modes of Teaching" OOFHEC2016 Proceedings, pp. 698–704 (2016)
16. Ferrara, V., Macchia, A., Sapia, S., Lella, F.: Cultural Heritage Open Data to Develop an Educational Framework, pp. 166–170 (2014)
17. Oomen, J., Aroyo, L.: Crowdsourcing in the Cultural Heritage Domain: Opportunities and Challenges, pp. 138–149 (2011)
18. Ridge, M.: From tagging to theorizing: deepening engagement with cultural heritage through crowdsourcing. Curator: Museum J. **56**(4), 435–450 (2013)
19. Wilkinson, M.D., Dumontier, M., Aalbersberg, I.J.J., Appleton, G., Axton, M., Baak, A., Blomberg, N., Boiten, J.-W., da Silva Santos, L.B., Bourne, P.E.: The FAIR guiding principles for scientific data management and stewardship. Sci. Data **3**(1), 1–9 (2016)
20. Nikiforova, A., McBride, K.: Open government data portal usability: a user-centred usability analysis of 41 open government data portals. Telematics Inform. **58**, 101539 (2021)
21. Jeeser, K.: Documentation of Museum Objects in Estonian Museums: Development and Application of Museum Information System (2009)
22. Europeana Foundation: Estonia and Europeana: An Overview (2021). https://pro.europeana.eu/files/Europeana_Professional/Europeana_Foundation_Governance/Member_States/Country_Reports/CountryReport_Estonia_January2021.pdf
23. Ioannides, M., Chatzigrigoriou, P., Bokolas, V., Nikolakopoulou, V., Athanasiou, V., Papageorgiou, E., Leventis, G., Sovis, C.: Educational Creative Use and Reuse of Digital Cultural Heritage Data for Cypriot UNESCO Monuments, pp. 891–901 (2016)

24. Open Knowledge Foundation: Open Definition 2.1 (2015). http://opendefinition.org/od/2.1/en/
25. OpenGLAM Foundation: OpenGLAM Principles (2018). https://openglam.org/principles/
26. Poole, N., Racine, B., Cousins, J.: We Transform the World with Culture: Europeana Strategy
    2015–2020. Policy Report. Europeana Foundation, The Hague (2014). https://Pro.Europeana.
    Eu/Files/Europeana_Professional/Publications/EuropeanaStrategy2020.Pdf

Part II
Supportive Technologies and Tools
for Smart Learning

Chapter 5
Discoverability of OER: The Case of Language OER

Maria Perifanou⊕ and **Anastasios A. Economides**⊕

Abstract One of the obstacles that prevent the widespread adoption of Open Educational Resources (OER) is the difficulty to find appropriate OER for specific educational objectives. This paper investigates this discoverability problem by searching for OER in eleven well-known repositories of OER (ROER). The search for "Language Game" and "Italian Language" OER was used as a case study. The search found very few useful language OER in these ROER. Also, it revealed a number of obstacles in finding appropriate OER such as the absence of a uniform structure of ROER, the absence of a uniform OER metadata description, inaccurate, obsolete, and missing metadata descriptions of OER, obsolete OER, not really open and free educational resources, and more. Finally, the paper makes suggestions for improving both ROER and OER description.

Keywords Find OER · Language OER · OER · OER repositories · Open educational resources · ROER · Search OER

5.1 Introduction

During the COVID-19 pandemic, traditional education was heavily disrupted by the schools' closure. Face-to-face and blended teaching were shifted to exclusively online teaching, and all educational activities were performed online. However, teachers and students were not prepared for online teaching and learning. Furthermore, many students did not have access to textbooks or laboratories. In such a challenging and urgent situation, many researchers and organizations [1, 2] suggested the use of Open Educational Resources (OER) by teachers and students. OER are learning, teaching, and research materials in any format and medium that reside in the public domain or are under copyright that has been released under an open license that permits no-cost access, use, adaptation, and redistribution by others [3]. Another definition requires

M. Perifanou (✉) · A. A. Economides
SMILE Lab, University of Macedonia, 54636 Thessaloniki, Greece

A. A. Economides
e-mail: economid@uom.gr

Ó. Mealha et al. (eds.), *Ludic, Co-design and Tools Supporting Smart Learning Ecosystems and Smart Education*, Smart Innovation, Systems and Technologies 249,
https://doi.org/10.1007/978-981-16-3930-2_5

the OER to meet the "5Rs Framework" [4] so that users are free to: (1) Retain: Users have the right to make, archive, and "own" copies of the content; (2) Reuse: Content can be reused in its unaltered form; (3) Revise: Content can be adapted, adjusted, modified, or altered; (4) Remix: The original or revised content can be combined with other content to create something new; and (5) Redistribute: Copies of the content can be shared with others in its original, revised, or remixed form. Recent definitions further extended the OER abilities to include the following: freely and openly find, access, store, use, create, interact, collaborate, evaluate, share, and abandon OER without any cost, at any place and any time [5].

OER is promoted by UNESCO [3] as a mean to meet Sustainable Development Goal 4 (SDG4) of the United Nations 2030 Agenda [6]. According to SDG4, all girls and boys should complete free, equitable, and quality primary and secondary education. However, most students cannot have access to textbooks due to their high cost [7–9]. Parents' and students' inability to buy textbooks affects negatively students' learning as well as access to courses and graduation. UNESCO [3] trusts that OER can support quality education that is equitable, inclusive, open and participatory. In addition, OER can provide teachers with a wider availability of educational material to choose for teaching and learning [3]. UNESCO [3] recommended five areas of actions to facilitate OER adoption: (i) Capacity building, capacity of education stakeholders to create access, reuse, adapt, and redistribute OER; (ii) Developing supportive policy; (iii) Effective, inclusive, and equitable access to quality OER; (iv) Nurturing the creation of sustainability models for OER; and (v) Facilitating international cooperation.

Similarly, European Commission [10] well recognized the value of OER to reduce the costs of educational materials and provide content adapted to the learners' needs. According to Timisoara Declaration [11], education should be open to all. This openness includes open access to digital resources and infrastructure as well as open access to people and professionals. It can be supported by OER. Timisoara Declaration was signed by Association for Smart Learning Ecosystems and Regional Development (ASLERD), European Association of Distance Teaching Universities (EADTU), European Association of Technology Enhanced Learning (EATEL), European Distance and E-learning Network (EDEN), International Association for e-Science (IAFeS). Furthermore, ICDE OER Advocacy Committee [12] recommends to support research and evaluation regarding OER work.

OER not only can save money [13] but they can also increase students' independence, self-reliance, satisfaction, and interest in the subjects taught as well as collaboration amongst learners and among teachers [14–17]. Furthermore, open textbooks could be of higher quality than copyright-restricted textbooks [18].

However, OER are yet to become mainstream due to various obstacles. These obstacles that prohibit OER broad adoption and use originate from the following sources: (i) ROER-based obstacles, (ii) OER-based obstacles, and (iii) User-based obstacles. The following list presents the specific obstacles:

ROER-based obstacles:

- difficulty to find OER and ROER [15, 19–28];

- cost of developing and maintaining effective interoperable ROER [29, 30].

OER-based obstacles:

- low-quality OER with respect to content timeliness, fit for purpose, usability, presentation, editing, multimedia, and more [28, 31, 32];
- limited availability of appropriate and suitable OER for various subjects and levels [10, 20, 21, 33];
- limited availability of OER in other languages beyond English [10, 34];
- OER incompatibility to local situation with respect to language, context, syllabus, students, etc. [33, 35–37];
- difficulty to adapt the OER to local situation [26, 38, 39];
- difficulty to sustain and keep on the existing OER updated and current [32];
- unclear or wrong copyrights of OER [20];
- difficulty to create an OER by combining and mixing OER that hold different copyrights and licenses [1];
- high cost of developing effective OER [34];
- lack of OER reviews, evaluations, quality assurance, and certifications [19, 40];
- limited use of universally accepted educational metadata standards for the OER [41].

User-based obstacles:

- lack of awareness of OER [20–22, 36, 37, 42];
- lack of digital infrastructure and tools to use OER [34];
- lack of digital skills, OER skills, and open licensing skills [25, 26, 34, 43–45];
- difficulty to integrate OER in class [19, 22, 23, 26, 37, 43, 46–49];
- lack of teachers' support from the educational institutes with respect to the absence of OER training, copyrights, rewards, and compensation, etc. [33, 37, 48];
- attachment to the ordinary and unwillingness to try new things [8].

More specifically, regarding the OER discoverability obstacle, several previous studies underlined the difficulty to discover appropriate OER to satisfy specific educational goals [15, 19–28]. Even European Commission [10] alerted that it is difficult to find adequate OER, and there is a need to make them more visible and accessible by all people.

More than half of 420 OER users in Asia pointed out that their inability to find specific and relevant OER was a serious obstacle in using OER [19]. Similarly, about 50% of thousands US faculty had difficulties to find appropriate OER [20, 21, 28]. Likewise, more than 50% of thousands educators all over the world faced difficulties in finding suitable OER in their subject, knowing where to find OER, and finding high-quality OER [15, 23, 50]. In another survey, US faculty stated that their inability to find OER was the second most significant barrier of adopting OER [22]. Also, educators in India expressed their inability to find OER on topics of interest [26].

A review of 51 previous studies on OER found that discovering OER was problematic due to misapplication of effective metadata that would ease the OER discoverability and sharing [25]. Finally, the Commonwealth of Learning (COL) makes efforts to implement the OER discovery services of the Directory of Open Educational Resources (DOER) using open standards and technologies [27]. COL emphasized that institutional ROER should follow open interoperability standards.

This paper investigates the problem of finding appropriate language OER. The next section describes two cases of searching and finding language OER. Section 5.3 discusses the results while Sect. 5.4 concludes and suggests directions for future research.

5.2   Method

After an extensive search in websites of universities' libraries, in the Registry of Open Access Repository (ROAR, http://roar.eprints.org/), and in the Directory of Open Access Repositories (Open DOAR, http://v2.sherpa.ac.uk/opendoar/), we identified well-known ROER (that contain language OER) and Directories of OER (that contain links to language OER). Next, we will use the term ROER to mean both ROER and Directory of OER. Major ROER that also curate language OER include the following in alphabetical order:

COERLL (https://www.coerll.utexas.edu/coerll/materials): The Center for Open Educational Resources and Language Learning (COERLL)'s mission is to produce and disseminate language OER such as online language courses, reference grammars, assessment tools, and corpora in more than twenty languages. Many of these OER have been evaluated and received a perfect score. It is the most well-known ROER that hosts and develops OER exclusively for foreign language teaching and learning.

Curriki (https://library.curriki.org/): It mainly concerns K-12 USA education. It offers OER for curriculum, lesson plans, apps, e-books, full courses, games, slides, video, etc. It is a K-12 global community for teachers, students, and parents to create, share, and find OER that improve teacher effectiveness and student outcomes.

DOER (Directory of Open Educational Resources) (http://doer.col.org/): The Commonwealth of Learning's (COL) designed DOER to help educators to find OER in the major ROER. It contains 114 language resources. More specifically, it contains links to 61 resources on language comprehension, 27 on language expression, 11 on grammar, 13 on vocabulary, etc.

Merlot (https://www.merlot.org/merlot/WorldLanguages.htm): It is a free and open online community of resources designed primarily for faculty, staff and students of higher education from around the world to share their learning materials and pedagogy. A lot of its OER are peer-reviewed online learning materials, cataloged by registered members and experts. Also, most of its OER are Creative Commons licensed. It contains information for over 3,000 World Languages materials including 165 language courses. Specifically, it provides the following: 271 World Languages Open Textbooks; 130 Textbook + English Language Arts; 45 ELL + Language, Grammar and Vocabulary; 27 Language Education (ESL) + Textbook; 23 ESL + Textbook; 20 Language Instruction + Textbook; 36 Languages Textbook.

MIT OpenCourseWare (OCW) (https://ocw.mit.edu/index.htm): MIT OCW provides openly almost all MIT course content. There are at least 79 language courses. MIT Global Languages (21G) offers language classes that cover nine different global language and culture groups. It includes more than 130 courses not only on language but also on arts and culture.

MOM (Mason OER Metafinder) (https://oer.deepwebaccess.com/oer/desktop/en/search.html): It is a search engine that searches 22 different sources such as the well-known ROER Merlot, OER Commons, OpenStax. Regarding foreign language, it provides 1120 results including 519 books and 27 e-books.

OASIS (https://oasis.geneseo.edu/): COL developed Openly Available Sources Integrated Search (OASIS) to facilitate the discovery of OER in different media formats. OASIS currently searches open content from 117 different databases and contains 388,707 records in a variety of subjects. It provides 823 language OER.

OER Commons (https://www.oercommons.org/): It contains over 50,000 OER in various languages and format: Audio, Braille/BR, Downloadable docs, eBook, Graphics/Photos, Interactive, Mobile, Text/HTML, and Video. There also curated OER Commons Collections on specific subjects as well OER Commons Hubs (network of users who can create and share collections, administer groups, and share news and events associated with a project or organization). Regarding language OER, it provides 1725 OER including 36 Language Instruction; 38 Language + Full Course; 13 Language Education (ESL) + Full Course; 1158 CCSS (Common Core State Standards) Aligned English Language Arts collections.

OpenStax (https://openstax.org/): It has a large collection of peer-reviewed open textbooks on a large variety of subjects. These textbooks are being used in over half of US college and universities and in over 100 countries. It has published 42 titles for college and high school courses across science, math, social sciences, business, and the humanities disciplines since 2012. More than 36.00 instructors worldwide have adopted an OpenStax textbook, and 14 million students have exploited its textbooks. There are not any language textbooks.

Open Textbook Library (https://open.umn.edu/opentextbooks/subjects/languages): It offers more than 700 peer-reviewed open textbooks to be freely used, adapted, and distributed. It provides 48 language textbooks.

OpenLearn (https://www.open.edu/openlearn/): The Open University supports OpenLearn which offers over 1000 free courses, interactives, quizzes, videos, and audios, etc. It includes 75 language courses with 60 language e-books and 57 language videos.

At a first glance, it seems that there are various alternatives to find appropriate OER for teaching and learning a specific subject. However, the reality is far away from this wish. This paper will further investigate the availability of OER for teaching and learning a foreign language. It looks that some ROER provide access to a variety of language OER (e.g., 488 OER for teaching and learning Spanish in Merlot [51]). However, searching for a specific language (even a popular language, e.g., Italian), a specific proficiency level, a specific students' age, a specific learning objective (e.g., pronunciation, grammar), and a specific type (e.g., video), it is extremely difficult to find any suitable OER. Even searching many well-known ROER with every possible combination of keywords or even looking OER by OER in these ROER, there are not a lot of results.

In order to examine in a planned way, the availability and the easiness of finding language OER in these well-known ROER, we designed two experiments. In the first case, we tried to find "Language Game" OER. In the second case, we tried to find "Italian Language" OER. The results are depicted in Table 5.1.

Although the meta-search engines of Merlot and MOM found many results, most of them are not relevant to the search criteria or exhibit other problems that will be discussed in the next section. The rest ROER present none or very few OER. If we also put some extra filters with respect to students' age, learning objective (e.g., pronunciation), media (e.g., video), etc., then it is extremely difficult to find any suitable OER. In the next section, we further discuss the problems faced in searching these ROER. We also make recommendations for improving both the structure of the ROER and the metadata description of the OER.

Table 5.1 Discovering "language game" and "Italian language" in ROER

ROER	Language games OER	Italian language OER
COERLL	–	3
Curriki	16	3
DOER	2	8
Merlot	131 (World Languages + Simulation) 421 (World Language + "Drill and Practice")	51 (World Languages/Italian materials) 42 ("Italian language" materials)
MIT OCW	–	1 course
MOM	364	894
OASIS	2	5
OER Commons	14 ("Language Education (ESL)" + Game)	1 ("Language Education (ESL)" + Italian)
OpenStax	–	1
Open Textbook Library	–	3 open textbooks
OpenLearn	–	3 courses, 2 podcasts, 1 test

5.3 Discussion

In this section, we further describe problems faced in discovering OER and make suggestions for overcoming such problems. No matter how sophisticated the search engines in the ROER are, manual investigation of the search results is still needed. Many of the search results are irrelevant to the search keywords. Problems and suggestions are presented as follows:

1. **Huge number of different disconnected ROER**: All over the world, almost every educational institute and organization (national or international) has its own independent ROER to serve its potential users. All these thousands ROER are isolated and not interconnected. Suggestions: *Develop a meta-search engine that is able to both find and search simultaneously all these ROER;*

2. **Not a uniform structure and organization of all ROER**: Each ROER has different objectives, serves different users (e.g., age, subject/discipline, language, learning objectives, and needs), and organizes differently its OER. Almost every ROER has a different structure and organization. Suggestions: *Develop an open standard to describe the structure and organization of ROER that will be easy to be linked to other ROER and for meta-search engines to search it. Promote this standard and motivate ROER to adopt and apply it;*

3. **Not a common OER metadata description across all ROER:** Most ROERs do not adopt an open educational metadata standard to describe their OER. On the contrary, they describe their OER metadata in their own way. Although there are open educational metadata standards (e.g., LOM, SCORM, and IMS), not all ROER use them [41]. Even worse, repositories that use such standards to describe the OER, very often mix different metadata standards in the same repository [41]. Suggestions: *Develop an open educational metadata standard to describe OER that is both usable and effective. Promote this standard and motivate ROER to adopt and apply it;*

4. **ROER in different languages**: In every country, the educational institutes and organizations develop ROER in the local language. However, it is difficult to explore and search repositories in a language that you do not speak. Although contemporary browsers may provide automatic translation to your language, most menus titles cannot be translated. Furthermore, a meta-search engine should be able to search these multilingual ROER and find OER in any language. Suggestions: *Develop an open standard to describe the structure and organization of ROERs in any language that can be searched by a search engine in any language. Develop an open educational metadata standard to describe OER in any language that is both usable and effective. Promote these standards and motivate ROER to adopt and apply them. Develop a meta-search engine that is able to both find and search simultaneously all multilingual ROER in any language;*

5. **Inaccurate search results at ROER**: Many of the search results in ROER are not accurate. For example, searching for "language game," the results include computer languages (e.g., C++) or philosophical issues on language. Also,

another problem appears when a resource is composed from several components. In this case, an ROER or a meta-search engine may consider it either as a single OER or as many different OER. So, the number of search results is not always an accurate index regarding the quantity of resources in a topic. For example, a course may be composed from one or more textbooks, tenths of chapters, videos, lessons, texts, cases, examples, exercises, questions, assignments, quizzes, etc. So, the search results can show not only this single course but also tenths of related materials to this course. For example, the Mason OER Metafinder (MOM) provides 100 OER results to searching "Italian language" at the MIT OpenCourseware. However, these 100 OER correspond to just two OER courses at the MIT OpenCourseware. Suggestions: *Develop an open educational metadata standard to describe OER in any language that is both usable and effective. This standard should allow the exact description of the OER linked to its modules but without excessive workload for the people who curate it. However, it is a difficult problem to accurately describe the OER with the least effort. Promote this standard and motivate ROER to adopt and apply it*;

6. **Very few OER in many ROER**: Since most ROER serve specific local users and have limited budget and resources, they cannot curate a large number of OER. Suggestions: *Promote the idea of OER all over the world. Motivate, interconnect, and train teachers and others to develop OER and populate ROER. Promote the interconnection among all ROER and the adoption of open standards*;

7. **Inaccurate and incomplete metadata description about the OER**: The OER metadata are filled by people. In some cases, these people do not fill in some metadata or they fill inaccurate information. For example, an OER was described as a game, but it was a kind of quiz. In other cases, there is no information regarding copyrights (e.g., "Not specified" or "Unknown"). Suggestions: *Develop an open educational metadata standard to describe OER in any language that is both usable and effective. Train the people who catalog and curate OER in ROERs. Periodically reconsider and check the metadata of all OER in the repository;*

8. **"OER" which are not really open and free**: In many cases, although an OER was described as being open and free, in reality there was a cost to use it. More specifically,

   a. Some resources are no longer OER, e.g., Merlot lists as OER the MFL games (https://www.merlot.org/merlot/viewMaterial.htm?id=86589). However, this is now a company called español eXtra (https://www.espanolextra.com/), and it uses subscriptions as a revenue model;

   b. Some resources are free for a sample of introductory material (e.g., one lesson, one quiz, one short video), and the rest material has to be paid e.g., español eXtra (https://www.espanolextra.com/), transparent language (https://www.transparent.com/), and German language games (https://www.pimsleur.com/);

   c. Some resources are free for a trial or for a limited time, e.g., Mango languages (https://mangolanguages.com/);

d. Some resources are free but they contain advertisements, e.g., GameZone (https://www.english-online.org.uk/games/gamezone2.htm), 50Languages (https://www.50languages.com/), and ESLGames (http://www.eslgamesworld.com/);

e. Some resources are free but they require to give your name and email in order to send you the resource.

Suggestions: *Catalog only real OER. Periodically reconsider and check the openness of all linked OER in the Repository;*

9. **Obsolete OER (including links)**: In several cases, the content or/and the technology used to create the OER are obsolete. For example, several OER use flash technology. However, Adobe does not support any more Flash. Also, in several cases, the ROER provided links to OER that did not exist, e.g., Broken links, "domain for sale," "App Not Available," "This app is currently not available in your country or region," "Forbidden: You do not have permission to access this resource," "This site cannot be reached," "ERROR_DEFAULT_404 PAGE NOT FOUND." Such obsolete information to non-existent OER was even given for information that has been recently updated (e.g., the label "Date Modified in MERLOT" states a recent date), e.g., LinguaWeb (https://www.merlot.org/merlot/viewMaterial.htm?id=89572) has been updated on September 24, 2020 and points to http://www.linguaweb.com/?f which does not exist on February 24, 2021 ("This domain is for sale. You can buy it right now!"). Suggestions: *Periodically reconsider and check for any changes in the OER metadata and remove any obsolete OER from the repository.*

5.4 Conclusions

Despite the cost savings and other benefits that OER bring, their widespread adoption by educators and schools remains to be seen. One serious obstacle to OER becoming mainstream is the difficulty to discover quality OER that are appropriate for specific educational objectives. This paper presents the results of searching for language OER in well-known ROER. Then, it records the problems faced and makes suggestions for improving the ROER and the OER description.

Administrators of ROER worldwide would collaborate and agree to adopt a common standard for the structure and organization of their ROER as well as a common educational metadata standard for the OER. Also, they should collaborate, train, and motivate curators, authors (creators), teachers, and others to develop, evaluate, and catalog quality OER for various audience, languages, subjects, educational levels, material types, format, media, etc. Future research may investigate the discoverability of OER for other languages and other subjects (besides language OER).

References

1. Huang, R., Liu, D., Tlili, A., et al.: Guidance on Open Educational Practices During School Closures: Utilizing OER under COVID-19 Pandemic in Line with UNESCO OER Recommendation. Smart Learning Institute of Beijing Normal University, China (2020). https://iite.unesco.org/wp-content/uploads/2020/05/Guidance-on-Open-Educational-Practices-during-School-Closures-English-Version-V1_0.pdf. Last Accessed 3 Mar 2021
2. International Commission on the Futures of Education: Education in a Post-COVID World: Nine Ideas for Public Action. United Nations Educational, Scientific and Cultural Organization (UNESCO), 7, place de Fontenoy, 75352 Paris 07 SP, France (2020)
3. UNESCO: Recommendation on Open Educational Resources (OER). The General Conference of the United Nations Educational, Scientific and Cultural Organization (UNESCO), meeting in Paris from 12 to 27 November 2019. https://unesdoc.unesco.org/ark:/48223/pf0000373755/PDF/373755eng.pdf.multi.page. Last Accessed 3 Mar 2021
4. Wiley, D.: Defining the "Open" in Open Content and Open Educational Resources (2014). https://opencontent.org/definition/. Last Accessed 3 Mar 2021
5. Economides, A.A., Perifanou, M.: Dimensions of openness in MOOCs & OERs. In: EDULEARN2018 Proceedings, 10th International Conference on Education and New Learning Technologies, pp. 3684–3693. 2–4 July, Palma Spain, IATED Digital Library (2018). https://doi.org/10.21125/edulearn.2018.0942
6. United Nations: Transforming Our World: The 2030 Agenda for Sustainable Development (2015). https://sdgs.un.org/2030agenda. Last Accessed 3 Mar 2021
7. Donaldson, R., Opper, J., Shen, E.: Student Textbook and Course Materials Survey: Results and Findings. Florida Virtual Campus, Office of Distance Learning & Student Services (2018). https://www.oerknowledgecloud.org/record2630. Last Accessed 3 Mar 2021
8. Fischer, L., Belikov, O., Ikahihifo, T.K., Hilton, J., III., Wiley, D., Martin, M.T.: Academic librarians examination of university students' and faculty's perceptions of open educational resources. Open Praxis **12**(3), 399–415 (2020)
9. Jenkins, J.J., Sanchez, L.A., Schraedley, M.A.K., Hannans, J., Navick, N., Young, J.: Textbook broke: textbook affordability as a social justice issue. J. Interact. Media Educ. **1**, 1–13 (2020). https://doi.org/10.5334/jime.549
10. European Commission: Opening Up Education: Innovative teaching and learning for all through new technologies and open educational resources. COM, 654. https://eur-lex.europa.eu/legal-content/EN/TXT/PDF/?uri=CELEX%3A52013DC0654&amp%3Bamp%3Bfrom=EN (2013). Last Accessed 3 Mar 2021
11. ASLERD: Timisoara declaration: Better learning for a better world through people centred smart learning ecosystems, pp. 1–9. http://www.mifav.uniroma2.it/inevent/events/aslerd/docs/TIMISOARA_DECLARATION_F.pdf (2016). Last Accessed 3 Mar 2021
12. Ossiannilsson, E., Aydin, C.H., Wetzler, J.: Report on the Survey on the UNESCO OER Recommendation: Conclusions and suggestions. ICDE OER Advocacy Committee (2020)
13. Hilton, J.: Open educational resources and college textbook choices: a review of research on efficacy and perceptions. Educ. Technol. Res. Dev. **64**(4), 573–590 (2016). https://doi.org/10.1007/s11423-016-9434-9
14. Blyth, C.: Opening up FL education with open educational resources: the case of *Français interactif*. In: Rubio, F., Thoms, J. (eds.) Hybrid Language Teaching and Learning: Exploring Theoretical, Pedagogical and Curricular Issues, pp. 196–218. Heinle Thomson, Boston (2012)
15. de los Arcos, B., Farrow, R., Pitt, R., Weller, M., McAndrew, P.: Adapting the curriculum: how K-12 teachers perceive the role of open educational resources. J. Online Learn. Res. **2**(1), 23–40 (2016). https://www.learntechlib.org/primary/p/151664/. Last Accessed 3 Mar 2021
16. Rossomondo, A.: The Acceso project and FL graduate student professional development. In: Allen, H.W., Maxim, H.H. (Eds.) Educating the Future FL Professoriate for the 21st Century, pp. 128–148. Boston: Heinle Cengage (2011)
17. Sabadie, J., Muñoz, J., Punie, Y., Redecker, C., Vuorikari, R.: OER: a European policy perspective. J. Interact. Media Educ. 1–12 (2014). https://doi.org/10.5334/2014-05

18. Kimmons, R.: OER quality and adaptation in K-12: comparing teacher evaluations of copyright-restricted, open, and open/adapted textbooks. Int. Rev. Res. Open Distrib. Learn. **16**(5) (2015). https://doi.org/10.19173/irrodl.v16i5.2341

19. Abeywardena, I.S., Dhanarajan, G., Chan, C.S.: Searching and locating OER: barriers to the wider adoption of OER for teaching in Asia. In: Proceedings of the Regional Symposium on Open Educational Resources: An Asian Perspective on Policy and Practices, pp. 19–21 (2012)

20. Allen, I.E., Seaman, J.: Opening the Curriculum: Open Educational Resources in US Higher Education. Babson Survey Research Group, Pearson (2014)

21. Allen, I.E., Seaman, J.: Opening the Textbook: Educational Resources in U.S. Higher Education, 2015–16. Babson Survey Research Group, Oakland. http://www.onlinelearningsurvey.com/reports/openingthetextbook2016.pdf (2016). Last Accessed 3 Mar 2021

22. Belikov, O., Bodily, R.: Incentives and barriers to OER adoption: a qualitative analysis of faculty perceptions. Open Praxis **8**(3), 235–246 (2016). https://doi.org/10.5944/openpraxis.8.3.308

23. de los Arcos, B., Farrow, R., Pitt, R., Perryman, L-A., Weller, M., McAndrew, P.: OER Research Hub Data 2013–2015: Educators. OER Research Hub. http://oro.open.ac.uk/47931/ (2015). Last Accessed 3 Mar 2021

24. Hodgkinson-Williams, C., Arinto, P.B.: Adoption and Impact of OER in the Global South. African Minds, International Development Research Centre & Research on Open Educational Resources, Cape Town & Ottawa (2017). https://doi.org/10.5281/zenodo.1005330

25. Luo, T., Hostetler, K., Freeman, C., Stefaniak, J.: The power of open: benefits, barriers, and strategies for integration of open educational resources. Open Learn. J. Open Distance e-Learn. **35**(2), 140–158 (2020). https://doi.org/10.1080/02680513.2019.1677222

26. Mishra, S., Singh, A.: Higher education faculty attitude, motivation and perception of quality and barriers towards OER in India. In: Hodgkinson-Williams, C., Arinto, P.B. (eds.) Adoption and Impact of OER in the Global South, pp. 425–458 (2017). Retrieved from https://doi.org/10.5281/zenodo.602784

27. Muthu, M., Cheng, R.: Directory of Open Educational Resources (DOER): A Discovery Service Framework to Provide Structured Access to OERs. Pan-Commonwealth Forum, 9–12 September 2019, Edinburgh, Scotland (2019)

28. Seaman, J.E., Seaman, J.: Opening the Textbook: Educational Resources in U.S. Higher Education. Babson Survey Research Group. http://www.onlinelearningsurvey.com/reports/openingthetextbook2017.pdf (2017). Last Accessed 3 Mar 2021

29. Atkins, D.E., Brown, J. S., Hammond, A.L.: A Review of the Open Educational Resources (OER) Movement: Achievements, Challenges and New Opportunities. Report to the William and Flora Hewlett Foundation. http://hewlett.org/wp-content/uploads/2016/08/ReviewoftheOERMovement.pdf (2007). Accessed 5 May 2021

30. D'Antoni, S.: Open Educational Resources the Way Forward: Deliberations of an International Community of Interest [Report for UNESCO]. http://openaccess.uoc.edu/webapps/o2/bitstream/10609/7163/1/Antoni_OERTheWayForward_2008_eng.pdf (2008). Last Accessed 5 May 2021

31. Jung, E., Bauer, C., Heaps, A.: Higher education faculty perceptions of open textbook adoption. Int. Rev. Res. Open Distrib. Learn. **18**(4) (2017). https://doi.org/10.19173/irrodl.v18i4.3120

32. Martin, T., Kimmons, R.: Faculty members' lived experience with choosing open educational resources. Open Praxis **12**(1), 131–144 (2020). https://doi.org/10.5944/openpraxis.12.1.987

33. Hassall, C., Lewis, D.I.: Institutional and technological barriers to the use of open educational resources (OERs) in physiology and medical education. Adv. Physiol. Educ. **41**(1), 77–81 (2017)

34. Berti, M.: Open educational resources in higher education. Iss. Trends Educ. Technol. **6**(1) (2018). https://doi.org/10.2458/azu_itet_v6i1_berti

35. Karunanayaka, S., Naidu, S. (eds.): Dreamweaving Open Educational Practice. The Open University of Sri Lanka, Nawala, Nugegoda, Sri Lanka (2016)

36. Mtebe, J., Raisamo, R.: Challenges and instructors' intention to adopt and use open educational resources in higher education in Tanzania. Int. Rev. Res. Open Distrib. Learn. **15**(1) (2014)

37. Wang, T., Towey, D.: Open educational resource (OER) adoption in higher education: challenges and strategies. In: 2017 IEEE 6th International Conference on Teaching, Assessment, and Learning for Engineering (TALE), pp. 317–319. IEEE (2017)
38. Krelja Kurelovic, E.: Advantages and limitations of usage of open educational resources in small countries. Int. J. Res. Educ. Sci. (IJRES) **2**(1), 136–142 (2016). https://files.eric.ed.gov/fulltext/EJ1105180.pdf. Last Accessed 3 Mar 2021
39. Ng, R.Y.K., Ng, K.K., Liu, B.: An empirical study on the usefulness, effectiveness and practicability of vocational and professional education and training's (VPET) open educational resources (OER) in the hotel industry. In: Proceedings of the International Conference on Blended Learning, pp. 265–276. Springer, Cham (2019)
40. Dichev, C., Dicheva, D.: Is it time to change the OER repositories role? In: Proceedings of the 12th ACM/IEEE-CS Joint Conference on Digital Libraries, pp. 31–34 (2012)
41. Santos-Hermosa, G., Ferran-Ferrer, N., Abadal, E.: Repositories of open educational resources: an assessment of reuse and educational aspects. Int. Rev. Res. Open Distrib. Learn. **18**(5), 84+ (2017). https://doi.org/10.19173/irrodl.v18i5.3063
42. Olcott, D.: OER perspectives: emerging issues for universities. Distance Educ. **33**(2), 283–290 (2012)
43. Guo, Y., Zhang, M., Bonk, C.J., Li, Y.: Chinese faculty members' Open Educational Resources (OER) usage status and the barriers to OER development and usage. Int. J. Emerg. Technol. Learn. (iJET) **10**(5), 59 (2015). https://doi.org/10.3991/ijet.v10i5.4819
44. Perifanou, M., Economides, A.A.: Designing teachers' training on adopting OERs in their teaching. In: Proceedings of the International Conference on Education and New Developments (END Conference), 26–28 June 2021, END Publ. (2021)
45. Kosmas, P., Parmaxi, A., Perifanou, M., Economides, A.A.: Open educational resources for language education: a review of existing guidelines for the creation, share, evaluation and use of Open Educational Resources. In: Proceedings of the 23rd International Conference on Human-Computer Interaction (HCII), 24–29 July, Washington DC, Springer (2021)
46. Coleman-Prisco, V.: Factors influencing faculty innovation and adoption of Open Educational Resources in United States Higher Education. Int. J. Educ. Hum. Dev. **3**(4), 1–12 (2017), http://ijehd.cgrd.org/images/vol3no4/1.pdf. Last Accessed 3 Mar 2021
47. Collis, B., Strijker, A.: Re-usable learning objects in context. Int. J. E-Learn. **2**(4), 5–16 (2003)
48. Jhangiani, R.S., Pitt, R., Hendricks, C., Key, J., Lalonde, C.: Exploring faculty use of open educational resources at British Columbia post-secondary institutions. BCcampus Research Report. BCcampus, Victoria, BC (2016)
49. Towey, D., Reisman, S., Chan, H., Demartini, C., Tovar, E., Margaria, T.: OER: Six perspectives on global misconceptions and challenges. In: 2019 IEEE International Conference on Engineering, Technology and Education (TALE), pp. 1–7. IEEE (2019)
50. de los Arcos, B., Weller, M.: A tale of two globes: exploring the north/south divide in engagement with open educational resources. In: Schopfel, J., Herb, U. (eds.) Open Divide: Critical Studies on Open Access, pp. 147–155. Litwin Books, Sacramento, CA (2018)
51. Perifanou, M., Economides, A.A.: Language OER in open repositories. In: Proceedings of EUNIS 2021—A New Era of Digital Transformation: Challenges for Higher Education. European University Information Systems organization, 9–11 June (2021)

Chapter 6
Dialogism Meets Language Models for Evaluating Involvement in CSCL Conversations

Maria-Dorinela Dascalu, Stefan Ruseti, Mihai Dascalu, Danielle S. McNamara, and Stefan Trausan-Matu

Abstract The use of technology as a facilitator in learning environments has become increasingly prevalent with the global pandemic caused by COVID-19. As such, computer-supported collaborative learning (CSCL) gains a wider adoption in contrast to traditional learning methods. At the same time, the need for automated tools capable of assessing and stimulating collaboration between participants has become more stringent, as human monitoring of the increasing volume of conversations becomes overwhelming. This paper introduces a method grounded in dialogism for evaluating students' involvement in chat conversations based on semantic chains computed using language models. These semantic chains reflect emergent voices from dialogism that span and interact throughout the conversation. Our integrated method uses contextual information captured by BERT transformer models to identify links in a chain that connects semantically related concepts from a voice uttered by one or more participants. Two types of visualizations were generated to depict the longitudinal propagation and the transversal inter-animation of voices within the conversation. In addition, a list of handcrafted features derived from the constructed chains and computed for each participant is introduced. Several machine learning algorithms were tested using these features to evaluate the extent to which semantic chains are predictive of student involvement in chat conversations.

M.-D. Dascalu · S. Ruseti · M. Dascalu (✉) · S. Trausan-Matu
University Politehnica of Bucharest, 313 Splaiul Independentei, 060042 Bucharest, Romania
e-mail: mihai.dascalu@upb.ro

M.-D. Dascalu
e-mail: dorinela.dascalu@upb.ro

S. Ruseti
e-mail: stefan.ruseti@upb.ro

S. Trausan-Matu
e-mail: stefan.trausan@upb.ro

M. Dascalu · S. Trausan-Matu
Academy of Romanian Scientists, Str. Ilfov, Nr. 3, 050044 Bucharest, Romania

D. S. McNamara
Department of Psychology, Arizona State University, PO Box 871104, Tempe, AZ, USA
e-mail: dsmcnama@asu.edu

© The Author(s), under exclusive license to Springer Nature Singapore Pte Ltd. 2022
Ó. Mealha et al. (eds.), *Ludic, Co-design and Tools Supporting Smart Learning Ecosystems and Smart Education*, Smart Innovation, Systems and Technologies 249,
https://doi.org/10.1007/978-981-16-3930-2_6

67

Keywords Dialogism · Computer-supported collaborative learning · Semantic chains · Language models

6.1 Introduction

Smart learning environments greatly benefit from the development of technology [1], by improving educational processes, reducing the time to perform certain tasks, increasing the availability of resources, and providing an ecosystem that stimulates creativity and the desire to learn. Thus, learning becomes easily accessible to all people from different parts of the world, and communication with from different cultures is at hand, while resources are readily available. Nevertheless, technology should remain an enabler, given a *people in place-centered* perspective [2] in which attractiveness, coupled with the adhesion, must transcend toward a sense of belonging. The need of such learning environments has dramatically increased during the COVID-19 pandemic [3]. The transition from physical classes to fully online environments was drastic and adaptation was required in a short amount of time; as such, smart learning environments eased this transition, while supporting both students and teachers. Since everything moved online, face-to-face class discussions transitioned to forums, chats, and video meetings.

In conjunction with the adoption of learning environments, computer-supported collaborative learning (CSCL) has become increasingly used in educational contexts due to its synergic effects among peers. Learners share their ideas and opinions, learn from each other, while having access to a wide range of materials. Chats and forums, the most commonly used CSCL environments, offer learners the opportunity to work together to solve problems and ask for help when encountering issues— more generally, students collaboratively build knowledge and share it among all participants [4]. Collective and individual learning processes intertwine one with another to create collaborative knowledge, which is spread among all participants [5].

In CSCL environments, technology empowers communication and collaboration throughout the learning process. However, from tutor's perspective, analyzing the resulting conversations is a time-consuming task due to their increased volume. Therefore, automated tools that analyze conversations and evaluate collaboration between participants have become a necessity. Moreover, collaboration and creativity among peers can be stimulated using automated processes [6].

Dialogism was first introduced by Bakhtin [7] as a philosophical theory focusing on the idea that everything is a continuous exchange and interaction between several voices [8]. From a dialogical perspective, discourse is modeled as a weave of interactions in natural language among people, with the essential goal of building meaning and understanding. Voices represent different points of view that spread throughout the discourse and influence it. Closely correlated with the concept of voice are multivocality and polyphony, which are key features in dialogism. Dialogism is considered a paradigm for CSCL [9], where voices (points of view) take the form of concepts or events that are propagated throughout the conversation by participants who share convergent or divergent perspectives [10]. Multivocality and polyphony are key features in dialogism and are closely correlated to the concept of voice.

Voices found in participants' contributions from CSCL conversations interact one with another and influence each other. Throughout a conversation, the inter-animation of the voices is a key component for the success of the collaboration. The interactions between the participants are reflected in their voices; therefore, polyphony can be an indicator of collaboration [11]. The evolution of voices throughout a conversation and their influence on other participants provide valuable insights into collaboration.

This paper introduces a method grounded in dialogism to evaluate students' involvement in chat conversations based on semantic chains identified using state-of-the-art language models, namely bidirectional encoder representations from transformers [BERT; 12].

The paper is structured as follows. The next section presents state-of-the-art methods and solutions. The third section introduces our method, the corpus used for evaluation, and the features derived from the identified semantic chains. The fourth section describes our results, together with the longitudinal and transversal visualizations of semantic chains (i.e., voices), followed by conclusions and future work.

6.2   State of the Art

The number of applications and plugins aimed to support learning processes is constantly growing, while the COVID-19 pandemic has further evidenced the power of technology in the learning processes. Due to the COVID-19 pandemic, most learning institutions had to operate entirely online. This would have been almost impossible 20 years ago due to the lack of technologies and auxiliary solutions. Smart learning environments, online communication platforms (Zoom, Microsoft Teams, Google Meet), and discussion channels (forums, chats) facilitated this transition. However, online collaboration is different from face-to-face discussions in a classroom. In online environments, emotional aspects of the conversation, reflected in facial expressions that a teacher can interpret and guide, are lost or at least diminished when video is enabled. Students lose focus more easily and jump from one topic to another, for example, by simply posting a picture or a link. Thus, following the ways in which students collaborate and maintain focus presents a far more tedious task for teachers.

The information that spreads during a conversation, the topics that are discussed and the immediate transitions from one topic to another, is key components in evaluating the collaboration between participants. CSCL environments aim to support students in the learning process by emphasizing collaboration using chats or forums. Dialogism, considered the theoretical framework of CSCL [9], was first introduced by Bakhtin [7] as a philosophical theory. According to Bakhtin, everything around us is a continuous change and an interaction between several voices. Multiple elements can be derived from the inter-animation of voices: convergence of points of view, potential conflicts, the way the information spreads, the change of a topic, or the taking over of another; all these elements lead to true polyphony [8].

Starting from Bakhtin's theory of polyphony and inter-animation of voices, Trausan-Matu et al. [13–15] proposed the polyphonic model of discourse analysis for chat conversations. The polyphonic model is based on the identification of voices and builds a graph-based representation of the conversation in which the inter-animation of voices generates convergences or divergences of points of view. Tracking events in a conversation in chronological order reflects a *longitudinal* dimension of discourse. The voices that propagate and inter-animate one with another reflect the interactions between the participants and their collaboration [11]. The exchange of ideas, the abandonment of one topic and the taking over of another, the divergences and convergences that may appear at a certain moment reflect a *transversal* dimension of the discourse [16].

From a computational point of view, voices represent semantic chains [17], which in turn can be generated using lexical chains [18] (i.e., sequences of words that are repeated or semantically related, including synonyms, hyponyms, or siblings). Jayarajan, Deodhare, and Ravindran [19] rely on nouns and compound nouns in identifying lexical chains. Traditional methods of identifying lexical relationships between words are based on Wordnet [20] or Roget's Thesaurus [21]. Mukherjee, Leroy, and Kauchak [18] used lexical chains identified from word repetitions, synonyms, and semantic relationships between nouns, to classify medical texts into two categories: easy and difficult. Ruas et al. [22] combined lexical chains with word embeddings to extract semantic relationships between words. Migrating to CSCL environments, the evaluation of collaboration and interactions between participants can be performed based on voice overlap [23]. The ways in which voices are emitted and propagate throughout the conversation, and the interconnection of the exposed points of view is all indicative of collaboration [23].

6.3   Method

In this study, we extend the method proposed by Ruseti et al. [24] to identify semantic chains in chat conversation and derive corresponding features from these chains predictive of student involvement.

6.3.1   Corpus

Our analysis is performed on the same chat conversations processed in detail by Dascalu et al. [11, 25]. This corpus consists of 10 chats selected from a corpus of more than 100 conversations which were scored by 4 raters. The conversations took place between four to five undergraduate students studying computer-human inter-action who debated on the advantages and disadvantages of specific CSCL technolo-gies. The students had known each other since previous courses. During the conver-sations, each participant was an advocate of a technology and tried to convince the other participants of the advantages of their chosen technology. Afterward, all partic-ipants had to come up together with a new solution which incorporated the discussed advantages.

6.3.2   Building Semantic Chains

Our method uses contextual information captured by BERT [12] to identify the links in a semantic chain. BERT is a transformer-based deep neural network that uses a mechanism of attention, which learns the contextual relationships between words from a text. BERT builds contextual representations of words by stacking multi-head attention layers.

A dataset derived from the TASA corpus (http://lsa.colorado.edu/spaces.html) containing potential links between words in a given context was automatically gener-ated given simple rules, namely (a) repetitions of words with the same lemma; (b) synonymy, hypernymy, or sibling relationships using WordNet [20]; and (c) coref-erences identified using spaCy (https://spacy.io). This dataset was used to train a model on how to effectively combine the different attention heads from BERT using a multi-layer perceptron in order to identify generalized semantic links [24].

The previously trained model was then used to identify all potential semantic links in a conversation, while accounting for all pairs of words. Starting from these links, semantic chains are generated as connected components in the graph obtained by interconnecting all links exceeding an imposed similarity threshold. The entire procedure is described in detail by Ruseti et al. [24].

6.3.3   Semantic Chain Features

The previous semantic chains computed with BERT are used to assess the involve-ment of students within the conversation. The chains cannot be directly used to build a prediction model; as such, a list of features was defined based on the constructed chains. Part of the features are inspired from LEX-1 [26] and are generally appli-cable to lexical chains, whereas the remaining ones are chat specific. All handcrafted features (see Table 6.1) are computed for each participant from the conversation.

The chat-specific features are designed to capture the interaction between partici-pants based on the semantic chains that span across their contributions. Each semantic chain represents a distinct topic discussed; therefore, part of the features consider how chains are initiated by a participant, and how semantic chains are afterward continued by the same or other speakers. Even if a chain is only included in the

Table. 6.1 Feature description

Feature name	Description
General features	
Chains	Count (#) and ratio of chains (i.e., how many semantic chains are used by a specific user divided by the overall count of semantic chains present in the conversation)
Large chains	Count (#) and ratio of large chains (i.e., semantic chains containing more than 4 words)
Varied chains	Count (#) and ratio of varied chains (i.e., semantic chains with more than one different lemma)
Large and varied chains	Count (#) and ratio of chains that are both large and varied
Chat-specific features	
Initiated chains	Count (#) and ratio of chains initiated by the participant
Independent chains	Count (#) and ratio of chains with only one participant
Avg. participants	Average participants per chain
Avg. words	Average words per chain for each participant
Continuations	Count (#) of backward links from the current participant to another participant
Avg. continuation length	Average words in the conversation for each backward link between different participants

contributions of a single participant, it still remains relevant for measuring participation because it denotes active involvement. Additional metrics are also considered to account for collaboration besides mere chain counts, for example, word occurrences per chain belonging to a specific participant (denoting topic coverage among participants), as well as the delay in the conversation before continuing a given voice (quantifying a pause in terms of topic continuation, measured as words in-between two occurrences from the same semantic chain in two contributions pertaining to different participants).

6.4 Results

6.4.1 Semantic Chains Visualizations

Interactive visualizations were introduced to highlight both a longitudinal propagation of voices (see Fig. 6.1) and transversal overlap of semantic chains between participants (see Fig. 6.2). The views were developed using Angular 6 (https://angular.io), while the links between words were drawn using SVG. The same chat excerpt from Fig. 6.1 was analyzed by Dascalu et al. [11]. Words are colored according to the semantic chain two which they belong as well as the corresponding links. Each row represents an utterance, enriched with following details: identifier, timestamp,

Id	Time	Participant ID (technology)	Utterance
222	02:43	1 (blog)	wiki for documentation and faqs.
223	02:43	2 (forum)	and a forum for technical support .
224	02:43	3 (wave)	forum for technical support and maybe chat for live support wave for collaboration/brainstorming/ document sharing .
226	02:43	2 (forum)	chat for live support inside the company .
227	02:43	4 (wiki)	yes, live support is a good ideea .
228	02:44	5 (chat)	we could also use chat for meetings ....for peope who can't come to the meeting .
229	02:44	3 (wave)	wave is better for that you can share documents etc, view what the other person is typing etc.
231	02:45	3 (wave)	i think that about wraps it up.
232	02:45	5 (chat)	does it support live video feed?
233	02:45	4 (wiki)	wave have also support for audio or video conversation?

Fig. 6.1 Longitudinal view of semantic chains within a conversation

and participant identifier, which was incrementally generated for anonymization and followed by the supported technology by each participant. Four semantic chains were identified in the conversation segment shown in Fig. 6.1: 1—concepts related to documentation (e.g., "documents," "documentation"); 2—concepts related to text-centered CSCL technologies (e.g., "forum," "chat," "share," "view"); 3—concepts related to CSCL technologies that integrate video (e.g., "wave," "meetings," "video"); 4—concepts related to actions facilitated by technologies (e.g., "support"). The identified semantic chains reflect the theme of the conversations, more specifically the advantages and disadvantages of the CSCL technologies considered.

In contrast with the initial findings of Dascalu et al. [11], our method identifies more semantic chains with more related words—for example, new semantic chains—concepts related to documentation and actions; more related words—"meetings" and "video" are related to "wave"; chats and forum are now aggregated together as language models grasp their similarity. For the same part of conversation, the initial results [11] did not identify the semantic chains related to documentation (fuchsia

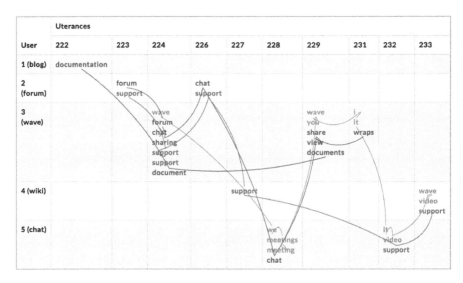

Fig. 6.2 Transversal view of semantic chains within a conversation

color) and actions (red color). Moreover, the initial results [11] did not include "meetings," "video," and "meetings" as related words to "wave," but the semantic chain contained only "wave." In addition, the new method identifies pronouns as part of the semantic chain (coreference resolution). Within the new method, "chat" and "forum" are now in the same semantic chain, while in the initial version [11], they were in separate chains. The sample denotes a higher cohesion between adjacent contributions, in contrast to previous results [11] in which a lower cohesion was argued for the same sample. Figure 6.2 presents a transversal view of the occurrences of semantic links whose underlying concepts are uttered by different participants. The words and the links are colored according to their corresponding semantic chain; colors differ between the visualizations because they are randomly selected.

6.4.2 Involvement Prediction

Several machine learning algorithms were tested using the previously introduced features to evaluate the performance of semantic chains in assessing involvement of students within the conversations. Since the dataset contained 10 distinct conversations, a tenfold cross-validation was performed, leaving one chat out for testing in each fold. Two different grades, namely participation (i.e., reflective of active involvement) and collaboration (i.e., interactions with peers) manually scored between 1 and 10, were predicted with two separate models. Table 6.2 shows a comparison of different models for the two tasks, and by taking into account the mean average error (MAE) on the two tasks, lower MAE values denote better models that are closer to

Table. 6.2 Model evaluation (values in bold mark the best performing model)

Model	Parameters	MAE participation	MAE collaboration
Inter-human agreement	–	0.907	0.819
Random forest	N trees = 10	**0.549**	0.655
Random forest	N trees = 100	0.551	**0.631**
SVR	RBF	0.662	0.674
SVR	Poly	1.391	1.397
Bayesian regression	–	0.625	0.658

the average human ratings. The human performance is approximated by computing the MAE of each rater compared to the average of all four ratings.

The human performance obtained on this dataset shows a MAE lower than 1 out of 10, which denotes a close, but not perfect agreement among raters (inherently, part were more relaxed, while the other were more fastidious); as such, 4 ratings were gathered for each participant and the system predicts the raters' average rating for both participation and collaboration. Random forest is the most predictive model reaching a MAE of 0.55; lowering the number of generated trees has a beneficial impact on performance, given the limited number of features and examples. Overall, the machine learning models seem to better capture the average scores of the raters, having the advantage of generalizing across all participants.

A subsequent analysis was performed to understand the importance of each feature for the two different tasks. The best model from the cross-validation for each task was trained on the whole dataset, and the Gini importance for each feature is presented in Table 6.3 [27]; higher values denote more relevant features, and the sum of all Gini feature importance scores is 1 for each prediction. The ratio of chains, denoting the coverage of semantic chains used by a participant in relation to the entire conversation, is by far the most predictive feature for participation with a score of 0.63. Continuations, reflective of links between two different participants (Gini importance of 0.35), coupled with counts and ratios of chains covered by the participant (Gini importance 0.16 and 0.15, respectively) are the best predictors for collaboration.

6.5 Conclusions and Future Work

CSCL environments help students learn by collaborating, sharing ideas and opinions, and finding the best solution together. CSCL technologies, such as chats and forums, are indispensable today, and they represent valuable sources for follow-up analyses.

Table. 6.3 Gini feature importance

Feature	Participation	Collaboration
# Chains	0.0047	0.1562
Chain ratio	**0.6294**	0.1457
# Large chains	0.0059	0.0235
Ratio of large chains	0.0064	0.0200
# Varied chains	0.0138	0.0457
Ratio of varied chains	0.0084	0.0631
# Large and varied chains	0.0024	0.0048
Ratio of large and varied chains	0.0275	0.0156
# Initiated chains	0.0036	0.0083
Ratio of initiated chains	0.0579	0.0274
# Independent chains	0.0034	0.0238
Ratio of independent chains	0.0881	0.0194
Avg. participants	0.0195	0.0262
Avg. words	0.0955	0.0307
# Continuations	0.0260	**0.3547**
Avg. continuation length	0.0076	0.0349

Bold denotes the most important feature for the considered predicted value

All activities moved online during the COVID-19 pandemic; as such, student assessment and monitoring become more challenging tasks for teachers. Collaboration between students in completing homework and their involvement in school activities are elements that must be taken into account in their assessment.

Within this paper, we introduce an automated method for evaluating students' involvement in chat conversations using dialogism as a paradigm and language models for identifying semantic chains. This study is an improvement on the analysis performed by Dascalu et al. [11]. Our method relies on contextual information captured by BERT to identify semantic links that are grouped into chains. Longitudinal and transversal visualizations were generated to highlight occurrence patterns of semantic chains, their propagation and co-occurrence throughout the conversation. In contrast to the previous results, our new method identifies more semantic chains with more related words, while also considering coreference resolution. A list of features was introduced, containing both general metrics applicable to the constructed semantic chains and chat-specific features, all computed for each participant. Several machine learning algorithms were tested using the defined features, and the best models based on random forest were capable to accurately predict participation and collaboration with a MAE of around 0.55 on a 10-point scale.

In terms of future work, we envision applying this dialogical model on other datasets and performing related analyses. For example, dialogism can be used to monitor student engagement in online courses, and the introduced features can be employed to predict dropout or course grades. Even more, by evaluating the introduced semantic chains, discussions and corresponding contributions can be cataloged as being course-specific, administrative, or off-topic, and an automated guidance mechanisms can be introduced, while targeting creativity stimulation.

Acknowledgements The work was funded by a grant of the Romanian National Authority for Scientific Research and Innovation, CNCS—UEFISCDI, project number TE 70 PN-III-P1-1.1-TE-2019-2209, ATES—"Automated Text Evaluation and Simplification." This research was also supported in part by the Institute of Education Sciences (R305A180144) and the Office of Naval Research (N00014-19-1-2424). The opinions expressed are those of the authors and do not represent views of the IES or ONR.

References

1. Chen, N.-S., Cheng, I.-L., Chew, S.W.: Evolution is not enough: revolutionizing current learning environments to smart learning environments. Int. J. Artif. Intell. Educ. **26**(2), 561–581 (2016)
2. Giovannella, C.: Smart learning eco-systems: "fashion" or "beef"? J. e-Learn. Knowl. Soc. **10**(3), 15–23 (2014)
3. Dhawan, S.: Online learning: a panacea in the time of COVID-19 crisis. J. Educ. Technol. Syst. **49**(1), 5–22 (2020)
4. Stahl, G.: Group Cognition. Computer Support for Building Collaborative Knowledge. MIT Press, Cambridge, MA (2006)
5. Cress, U.: Mass collaboration and learning. In: Luckin, R., Puntambekar, S., Goodyear, P., Grabowski, B., Underwood, J., Winters, N. (eds.) Handbook of Design in Educational Technology, pp. 416–424. Routledge, New York (2013)
6. Stamati, D., Dascalu, M., Trausan-Matu, S.: Creativity stimulation in chat conversations through morphological analysis. University Politehnica of Bucharest Scientific Bulletin Series C—Electr. Eng. Comput. Sci. **77**(4), 17–30 (2015)
7. Bakhtin, M.M.: The Dialogic Imagination: Four Essays. The University of Texas Press, Austin and London (1981)
8. Bakhtin, M.M.: Problems of Dostoevsky's Poetics. University of Minnesota Press, Minneapolis (1984)
9. Koschmann, T.: Toward a dialogic theory of learning: Bakhtin's contribution to understanding learning in settings of collaboration. In: International Conference on Computer Support for Collaborative Learning (CSCL'99), pp. 308–313. ISLS, Palo Alto (1999)
10. Trausan-Matu, S.: Automatic support for the analysis of online collaborative learning chat conversations. In: 3rd International Conference on Hybrid Learning, Vol. LNCS 6248, pp. 383–394. Springer, Beijing, (2010)
11. Dascalu, M., Trausan-Matu, S., McNamara, D.S., Dessus, P.: ReaderBench—automated evaluation of collaboration based on cohesion and dialogism. Int. J. Comput.-Support. Collab. Learn. **10**(4), 395–423 (2015)
12. Devlin, J., Chang, M.-W., Lee, K., Toutanova, K.: Bert: pre-training of deep bidirectional transformers for language understanding. arXiv preprint arXiv:1810.04805 (2018)
13. Trausan-Matu, S., Stahl, G., Zemel, A.: Polyphonic Inter-animation in Collaborative Problem Solving Chats. Drexel University, Philadelphia (2005)
14. Trausan-Matu, S., Dascalu, M., Rebedea, T.: PolyCAFe—automatic support for the polyphonic analysis of CSCL chats. Int. J. Comput.-Support. Collab. Learn. **9**(2), 127–156 (2014)
15. Trausan-Matu, S.: The polyphonic model of collaborative learning. In: Mercer, N., Wegerif, R., Major, L. (eds.) The Routledge International Handbook of Research on Dialogic Education, pp. 454–468. Routledge, Abingdon, UK (2020)

16. Trausan-Matu, S., Stahl, G., Sarmiento, J.: Supporting polyphonic collaborative learning. E-serv. J. Indiana Univ. Press **6**(1), 58–74 (2007)
17. Dascalu, M., Trausan-Matu, S., Dessus, P.: Voices' inter-animation detection with Reader-Bench—modelling and assessing polyphony in CSCL chats as voice synergy. In: 2nd International Workshop on Semantic and Collaborative Technologies for the Web, in conjunction with the 2nd International Conference on Systems and Computer Science (ICSCS), pp. 280–285. IEEE, Villeneuve d'Ascq, France (2013)
18. Mukherjee, P., Leroy, G., Kauchak, D.: Using lexical chains to identify text difficulty: a corpus statistics and classification study. IEEE J. Biomed. Health Inform. **23**(5), 2164–2173 (2018)
19. Jayarajan, D., Deodhare, D., Ravindran, B.: Lexical chains as document features. In: Proceedings of the Third International Joint Conference on Natural Language Processing: Volume-I (2008)
20. Miller, G.A.: WordNet: a lexical database for English. Commun. ACM **38**(11), 39–41 (1995)
21. Kipfer, B.A., Chapman, R.L.: The Concise Roget's International Thesaurus. HarperTorch (2003)
22. Ruas, T., Ferreira, C.H.P., Grosky, W., de França, F.O., de Medeiros, D.M.R.: Enhanced Word Embeddings Using Multi-Semantic Representation Through Lexical Chains. Information Sciences (2020)
23. Dascalu, M., Trausan-Matu, S., Dessus, P., McNamara, D.S.: Dialogism: A Framework for CSCL and a Signature of Collaboration. In: 11th International Conference on Computer-Supported Collaborative Learning (CSCL 2015), Vol. 1, pp. 86–93. ISLS, Gothenburg, Sweden (2015)
24. Ruseti, S., Dascalu, M.-D., Corlatescu, D.-G., Dascalu, M., Trausan-Matu, S., McNamara, D.S.: Exploring dialogism using language models. In: 22nd International Conference on Artificial Intelligence in Education (AIED 2021). Springer, Utrech, Netherlands (Online) (in press)
25. Dascalu, M., McNamara, D.S., Trausan-Matu, S., Allen, L.K.: Cohesion network analysis of CSCL participation. Behav. Res. Methods **50**(2), 604–619 (2018)
26. Somasundaran, S., Burstein, J., Chodorow, M.: Lexical chaining for measuring discourse coherence quality in test-taker essays. In: Proceedings of COLING 2014, the 25th International conference on computational linguistics: Technical papers, pp. 950–961 (2014)
27. Menze, B.H., Kelm, B.M., Masuch, R., Himmelreich, U., Bachert, P., Petrich, W., Hamprecht, F.A.: A comparison of random forest and its Gini importance with standard chemometric methods for the feature selection and classification of spectral data. BMC Bioinform. **10**(1), 1–16 (2009)

Chapter 7
Selfit—Accounting for Sexual Dimorphism in Personalized Motor Skills Learning

**Laurentiu-Marian Neagu, Eric Rigaud, Vincent Guarnieri,
Gabriel-Dănuț Matei, Sébastien Travadel, and Mihai Dascalu**

Abstract The development and the upkeep of psychomotor skills are fundamental for performing safely and efficiently daily life, professional, or leisure movements. Artificial intelligence offers the potential to develop adaptive systems that can support people's abilities to perform essential and specialized movements, by providing individualized and optimized learning sessions. Research on men is the building block of most theories and practices in the psychomotor development domain. However, it is assumed that women's inclusion will create potential interference due to their physiological variability (i.e., menstrual cycle and the effect of oral contraceptive). Thus, personalization of motor skills learning must consider significant differences between men's and women's physiology. Our study describes female-specific issues on psychomotor skills development (e.g., strength level, menstrual cycle, and female athlete triad). The sexual dimorphism-based individualization principles and parameters are afterward presented, together with their implementation in the *Selfit* intelligent tutoring system for psychomotor skills development. A risk injury monitoring module was implemented using computer vision technologies to track user movements and evaluate risk. A simulation with a virtual population of hundreds of men

L.-M. Neagu · E. Rigaud · V. Guarnieri · G.-D. Matei · S. Travadel
Centre of Research On Risks and Crisis Management, MINES ParisTech, PSL University, 1 Rue Claude Daunesse, Sophia Antipolis, Paris, France
e-mail: laurentiu.neagu@upb.ro

E. Rigaud
e-mail: eric.rigaud@mines-paristech.fr

V. Guarnieri
e-mail: vincent.guarnieri@mines-paristech.fr

G.-D. Matei
e-mail: gabriel-danut.matei@mines-paristech.fr

S. Travadel
e-mail: sebastien.travadel@mines-paristech.fr

L.-M. Neagu · G.-D. Matei · M. Dascalu (✉)
Computer Science Department, University Politehnica of Bucharest, 313 Splaiul Independentei, Bucharest, Romania
e-mail: mihai.dascalu@upb.ro

© The Author(s), under exclusive license to Springer Nature Singapore Pte Ltd. 2022
Ó. Mealha et al. (eds.), *Ludic, Co-design and Tools Supporting Smart Learning Ecosystems and Smart Education*, Smart Innovation, Systems and Technologies 249,
https://doi.org/10.1007/978-981-16-3930-2_7

and women was performed, demonstrating that considering sexual dimorphism in the tutoring module is critical for injury prevention and for providing personalized sessions for women training.

Keywords Motor skills learning · Training personalization · Psychomotor · Sexual dimorphism

7.1 Introduction

Sexual dimorphism accounts for the presence of differences between males and females, for example, in terms of size or body composition [1]. A common difference is that the human males exceed human females in size. Also, there is increasing evidence that body composition sex differences are observable even before puberty [2], and more evident when maturity is reached [1]. The aim of the current work is to present an updated version of *Selfit* [3], a smart ecosystem developed for training adult psychomotor skills [4], based on an intelligent tutoring system (ITS) architecture, which now considers the sexual dimorphism dimension. *Selfit* is designed for anatomical adaptation training, which is suitable for beginners and its main goal is to perform sport for health.

The updated system includes a monitoring module, required for the assessment of risk injury while training, as well as a novel tutoring approach based on a menstrual cycle-specific mesocycle calendar. Updates were also performed on core ITS components, namely the domain, student, and tutoring models. Domain-specific knowledge is modeled using ontologies [5] that establish a formal and explicit description of concepts and relations from the psychomotor skills domain.

The paper is structured as follows. The next section describes female-specific characteristics on psychomotor skills development, while presenting morphological and physiological differences, specific risks of injuries, and menstrual cycle. Afterward, the *Selfit* existing system is briefly presented, alongside integration details for the sexual dimorphism dimension; the updates for each of the impacted ITS modules are explained. The third section also presents an implementation of a computer vision-based system used to label body points, while movements are performed, in order to detect the risk of injury for the trainee. Next, the results section includes a simulation conducted to assess the effects of training women like men. A discussion based on the method and results is presented, while potential future research directions conclude the current work.

7.2 Related Work

Psychomotor skills development is a lifelong process of learning how to move accordingly to a dynamic environment. A movement competence is a transaction between an individual and a movement task within an environment. Essential movements, such as pushing, pulling, core, knee, or hip-dominant exercises are prerequisites for learning specialized, complex psychomotor tasks required by daily life, professional, or leisure activities. Learning to perform a movement safely and efficiently requires practicing an adequate volume of exercises for enhancing associated physical qualities, such as strength, flexibility, or endurance.

A recent systematic literature review on ITSs used for psychomotor skills development [6] has shown that there are several research groups working on building systems in several fields, from military to manual procedures or learning how to drive. Starting with this review, the *Selfit* ITS for psychomotor skills development is developed. The *Selfit* ITS aims to develop trainees' psychomotor skills, while performing fundamental and specialized movement tasks correctly and safely, through the generation of learning sessions adapted to their physical stimuli responses. First, the system defines the general learning objectives by interacting with trainees and deducing the acquired movement skills. Then, trainees perform fundamental upper and lower body movement tasks: ordered by incremental difficulty on a scale from 1 to 5. The updated version of the architecture can be seen in Fig. 7.1; the new submodules are marked in blue.

Evaluation of trainee performance provides the initial readiness to perform the movement skills profile. The development consists of twelve weeks of learning sessions, having between three and five sessions per week. Before each session, *Selfit* generates content using the initial training evaluation, while estimating the daily

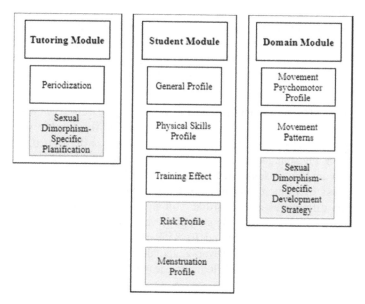

Fig. 7.1 *Selfit* ITS—updated architecture

training potential, the session length, and the trainee current step in the training plani-fication. According to the trainee's feedback after finishing the workout, *Selfit* updates the trainee parameters based on the subjective a posteriori assessment. Periodically, or at the end of the program, evaluation tasks validate the trainee's psychomotor skills progress—i.e., establish objective assessment. The work presented in this paper is related to the refinement of the *Selfit* system considering sexual dimorphism dimension.

7.3 Accounting for Sexual Dimorphism in the *Selfit* System

7.3.1 Basis of Human Sexual Dimorphism

The morphological, cognitive, and physiological differences between males and females impact the content development of the training sessions, together with asso-ciated risks of injuries and psychological disorders. This section briefly describes these differences.

Morphological differences. The difference between women's and men's skeleton size and body composition (i.e., density, relative fat mass, and lean body mass adjusted by height) varies at different age periods [1, 7, 8]. These differences increase at puberty due to hormonal differentiation. Adult human males are 7% taller than females, and there is a substantially higher amount of body fat and a substantially lower amount of lean body mass among women. Women have a smaller thorax, a larger abdomen, a broader and shallower pelvis, shorter legs, and a lower relative center of gravity than the male. The distribution of lower body muscle mass is quite similar between the sexes, whereas women have less muscle mass in the upper body than males.

Physiological differences. Human performance mainly relies on the anaerobic energy sector (effort duration from 1 to 100 s) and on the aerobic sector (effort duration from 100 s to several hours). The anaerobic sector depends on local stores of phosphagen. Phosphagen store per unit volume of muscle is independent of sex, but as women are less muscular than men, they are disadvantaged. Aerobic capacity is 30–40% less in women than in men due to body size, and blood hemoglobin concentration differences affect cardiac volume, blood volume, and peak oxygen transport. Females vary in absolute strength levels: between 40 and 60% of males' strength levels in the upper body and 70–85% in the lower body. The rate of force development (RFD) is lower in the average female than in the average male. Females generate less power than men, while having less power per unit of muscle volume [7, 8].

Cognitive and brain differences. Various theories [9–11] enunciate the difference between women's and men's cognitive abilities impacting psychomotor skills. In general, women tend to adapt their behavior to their perception of another person's emotions and thoughts. For navigation, they favor an egocentric strategy while using street names and building shapes as landmarks. They outperform males in precision and fine hand abilities, object location and verbal memory, verbal recognition, and semantic fluency tasks. In contrast, men tend to analyze and explore rules that govern a system. In general, they perform better on mental rotation and spatial navigation tasks than women. For navigation, they tend to favor an allocentric strategy which considers accurate judgments of distance. Men integrate speed and precision more quickly than women, and they tend to be better at sensorimotor tasks, including aiming, catching, and throwing.

The menstrual cycle. The menstrual cycle consists of the physiological phenomena which prepare the woman's body for possible fertilization. Menstruation occurs approximately every 24–35 days. The menstrual cycle comprises three phases: menses (or menstruation), follicular, and luteal. The process contains varying amounts of two endogenous hormones: estrogen (low during the menses, high during the follicular phase, and moderates during the luteal phase) and progesterone (low during menses and follicular phase, increased during the luteal phase). Hormone variations impact aerobic and anaerobic performances, while negatively affecting healthy body weight, bone density, overall responses to training, as well as increasing injury potential [7, 12, 13].

Female-specific risks of injuries. Women's morphological differences and hormonal variability during the menstrual cycle induce a higher risk of injury and, in particular, for anterior cruciate ligament (ACL) injury. The prevalence of ACL injury for women is 2–10 times greater than in males, for the same psychomotor activities. This prevalence causes a lower rate of force development, hamstring activation deficits, and greater ankle dorsiflexion, combined with the valgus position of the knees and external rotation of the hip [7, 14]. Moreover, three interdependent conditions predispose, when combined, female athletes in particular to a greater risk of illness and injury: **a** disordered eating (including anorexia nervosa and bulimia nervosa), **b** osteoporosis (loss of bone density), and **c** amenorrhea (lack of menstrual cycle).

7.3.2 Sexual Dimorphism for Scheduling Training Sessions

Accounting for sexual dimorphism when scheduling and designing learning sessions. Traditionally, the schedule and the design of psychomotor skills training sessions rely on physiological temporal variables. The delay between two sessions must be long enough not to induce overfatigue and not too long not to induce detraining. The evolution of session loads, computed using the Eq. 7.1, follows temporal patterns with progressive development from one week to another. The easiest week is set every three weeks to facilitate learning assimilation—for example, one week with easy sessions, one week with medium sessions, one week with heavy sessions, followed again by easy sessions.

Table 7.1 Training load adaptation to menstrual phase adapted from Pitchers [13]

Menstrual phase	Early follicular	Mid-follicular	Late follicular	Early luteal	Mid-luteal	Late luteal
Training load	Light	Medium	Medium /heavy	Very heavy	Medium	Light

$$\text{Session Load} = \sum \text{No Reps}_{\text{Ex } k} * \text{Intensity}_{\text{Ex } k} * \text{Rest Time}_{\text{Ex } k} * \text{No Joints}_{\text{Ex } k}$$
$$(7.1)$$

Trainee's performance monitoring and feedback support load adjustment. When scheduling learning sessions for women, tutors have to follow the same physiological temporal variables. However, when designing sessions, they must synchronize with the menstrual cycle to define sessions' content. Table 7.1 describes the relationship between session load and the different phases of the menstrual cycle according to Pitchers [13].

Accounting for sexual dimorphism when profiling trainees. Trainee profiling before starting and during a psychomotor skill development program generally consists of assessing physical capacities and identifying areas of weakness or pain associated with performance [7]. The consideration of female-specific risks of injuries, particularly ACL and female athlete triad, requires integrating appropriate tests of assessing the injury susceptibility. Results from these tests are used afterward to provide dedicated prophylaxis sessions and adjust learning sessions accordingly.

7.3.3 Sexual Dimorphism Updates on the Selfit Architecture

Sexual Dimorphism updates on the *Selfit* Domain model. The *Selfit* domain model consists of an ontology describing psychomotor skills competencies [15]. The ontology supports the learning process by providing answers to requests related to (a) learning objectives definition, (b) trainee evaluation, (c) learning program definition, and d) adaptation to student characteristics. The ontology core consists of the following classes: movement skill, trainee psychomotor skills profile, movement patterns, and training program modalities. The ontology is completed to provide answers to request related to: Which session content is the most appropriate to the student, depending on her position in her menstrual cycle? What is the student risk status? and How the learning session has to be updated? Consequently, classes related to requirements for designing sessions based on different phases of the menstrual cycle and classes describing female-specific risks of injuries and associated tests and prophylaxis measures have been added to the initial ontology.

Sexual Dimorphism updates on the *Selfit* Student model. The *Selfit* student model contains information on the trainee's psychomotor skills capacities, with focus on the skills related to the super-compensation cycle status. Moreover, it includes system usage statistics. The monitoring module collects information on how the trainees are using the system, together with their daily progress, and correspondingly updates the training parameters. The *Selfit* student model supports the generation of training sessions and the monitoring of trainee efficiency to optimize progression, while ensuring motivation to practice. The sexual dimorphism update consists of the addition of two dimensions in the student model. The first one aims to describe the evolution of female-specific risks while training, and to introduce the appropriate mitigation tasks to be performed when a risk value exceeds an imposed threshold. The second dimension aims to integrate the evolution of women's menstrual cycle and to properly select training session loads based on the women's calendar.

Sexual Dimorphism updates on the *Selfit* Tutoring model. The *Selfit* tutoring model supports the learning process by providing machine learning mechanisms to support the learning program's adaptation to the trainee's characteristics. Sport training is a complex process, which supports adaptation and personalization, while considering different temporalities: exercise, session (30 min to 120 min), microcycle (one week), mesocycle (one-month), and macrocycle (three months). Starting with predefined templates for microcycles and sessions, a novice trainer generates the most appropriate learning task using a multiarmed bandit strategy [3]. The task is composed of exercise, level, number of repetitions, and number of sets, by considering student characteristics, history, and his current state. The trainee provides direct feedback in the system, specifying the number of repetitions in reserve (defined as the number of repetitions he could have performed until failure), rate of perceived effort, and fatigue perception. The ratio between the planned and the perceived effort supports the evaluation of the session effectiveness and the baseline for algorithm learning. The dynamic of charge (i.e., complexity of exercises, number of repetitions, number of sets, and load) follows a calendar with a set of training rules defined by sport coaches. The charge is catalogued as easy in the first week, then medium, heavy, followed again by easy. The sexual dimorphism update consists of a calendar differentiation considering the dynamic of charge for men versus women and integrating the menstrual cycle. The training session content and charge are defined based on the trainee menstrual phase.

7.3.4 Computer Vision-Based Injuries Risk Assessment Module

A women's ACL injury risk assessment module was developed to support student initial screening and injuries risks monitoring. The assessment process is structured with four phases (see Fig. 7.2). The first phase consists of capturing one frontal and one sagittal video of the student performing a back squat. In the second phase, a human motion recognition module provides a discretization of each body joints' trajectory performing the back squat. Then, the risk assessment module analyzes this

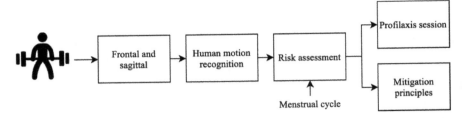

Fig. 7.2 ACL injury risk assessment module process

model for calculating the ACL injury risk factor. Finally, if risks are detected, the module provides instructions to the ITS.

Phase 1. The first phase consists of capturing one frontal and one sagittal video of the student performing a back squat. The back squat exercise allows identifying biomechanical deficits (inefficient motor unit coordination or recruitment, muscle weakness, strength asymmetry or joint instability, joint immobility, or muscle tightness) [13]. The two videos allow having a complete perspective of the performed movement and then provide information to assess knee valgus and ankle flexibility, two of the essential factors involved in the ACL injury risk. The two videos are necessary to have a complete perspective of the performed movement. For example, when analyzing the knee joint movement, the sagittal view allows calculating the angle between student's hip, knee, and ankle, and the frontal view gives information about the distance between student's knee and the vertical line passing through the middle of its ankle. The combination of these values supports the assessment of the knee valgus.

Phase 2. In the second phase, a human motion recognition module provides a discretization of each body joints' trajectory performing the back squat. The Open-Pose 2D pose estimation library [16, 17] allows obtaining videos with an added overlay containing the body key points and lines and a collection of JSON files, one for each frame, containing the position of the essential points, in pixels, as it can be seen in Fig. 7.3. A recent study [18] demonstrates the reliability and the validity of this library results by comparing them to results provided by a kinematic measurement by three-dimensional motion analysis devices using VICON.

Phase 3. The risk assessment module analyzes this model for calculating the ACL injury risk factor. The monitoring of the medial aspect of either knee passing the medial malleolus from the anterior perspective during any phase of the squat supports identifying a valgus [14]. The module evaluates the distance between "R or L Knee" and "R or L Ankle" when the student performs the back squat. If this distance is negative and greater than five centimeters, it means that the trainee has a valgus. The following figures describe a case where there is no valgus (see Fig. 7.4) and a case where there is a valgus on the left knee (see Fig. 7.5).

In the first case (cf. Fig. 7.4), the squat movement starts in frame 10 and finishes in frame 70. At the initial position, the distance between the knee and ankle is negative

"Neck"
"LShoulder"
"LElbow"

"MidHip"
"LHip"

"LKnee"
"LAnkle"
"LSmallToe"
"LHeel"

Fig. 7.3 ACL injury risk assessment module demonstration

Fig. 7.4 Illustration of the evolution of the distance between student's knee and ankle without valgus

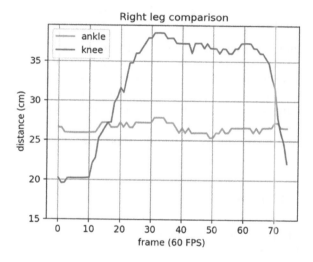

but less than 5 cm. As soon as the movement starts, the distance positively increases and then decreases until the movement is over. This is the behavior of a student without valgus.

In the second case (see Fig. 7.5), the squat movement starts in frame 30 and finishes in frame 150. At the initial position, the distance between the knees and ankles is negative, and the distance between knees and ankles is superior to 5 cm, demonstrating the presence of a valgus position before the start of the movement. During the movement, the negative distance between the left knee and the left ankle increases, demonstrating the valgus position's increase. On the right side, the distance decreases and becomes positive during the movement showing a correct position. At

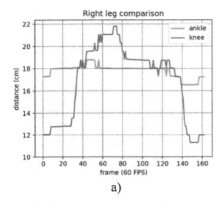 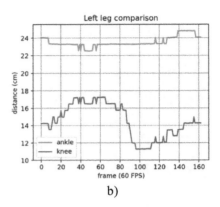

Fig. 7.5 Illustration of the evolution of the distance between student's knees and ankles with **a** a valgus on the left knee and **b** no valgus on the right one

the end of the movement, the distance between knees and ankles demonstrates a valgus position for both sides.

Phase 4. The risk of ACL injury increases if both a valgus and flexible ankles are detected. During the ovulation period of the menstrual cycle, hormones maximize the risk. If the risk assessment module detects a risk of ACL, it communicates with the *Selfit* ITS by providing mitigation rules to redesign the planned and future learning sessions for considering low-risk tasks. Furthermore, *Selfit* provides additional prophylactic learning sessions to decrease the risk of ACL with specific strength, flexibility, and proprioceptive tasks.

7.4   Results

The *Selfit* tutoring model supports the learning process by providing ML mechanisms to support the adaptation of the learning program to the trainee's characteristics. Sport training is a complex process, which supports adaptation and personalization while considering different temporalities (e.g., exercise, session, week, and month). The tutoring process is structured on a four-level maturity scale (i.e., novice, intermediate, advanced, and experts' trainers). The initial *Selfit* intelligent tutoring system used a novice trainer approach for tutoring adaptation. The novice trainer implements the multiarmed bandits' algorithm for personalizing the training sequences in a session. The intermediate trainer can personalize a session content; the advanced trainer personalizes the microcycle, mesocycle, and macrocycle content.

For the novice coach, the sessions were generated using a standard calendar plan, used by trainers in their daily work, without integrating the sexual dimorphism dimension. Current work presents the calendar split between men and women, and the plans

Table 7.2 Mesocycle beginner load pattern differences between men and women

Man load	L	L	L	L	L	L	L	M	M	M	M	M	M	M
Woman load	L	L	L	L	L	M	M	M	H	H	H	H	H	H
Day in month	1	2	3	4	5	6	7	8	9	10	11	12	13	14
Man load	H	H	H	H	H	H	H	L	L	L	L	L	L	L
Woman load	H	H	H	H	H	H	M	M	M	M	L	L	L	L
Day in month	15	16	17	18	19	20	21	22	23	24	25	26	27	28

are computed in mesocycles (4-weeks long). The 28-day beginner load pattern differences can be seen in Table 7.2, and the example includes a man starting first day of training and a woman starting training in the same day with her first early follicular (menses) day.

For the days of the month which are shown as light for man and medium for woman (e.g., days 6, 7 in month), if a woman would have been trained like a man, the session is considered suboptimal. A sequence of suboptimal sessions would produce a suboptimal training. This will imply more time to get the expected results and also the risk of losing trainee motivation, sessions may be too easy. On the other side, e.g., for day 21, a man would get a heavy session, while an optimal one for the woman would be medium—based on the cycle day, it is her first mid-luteal day. If the woman is trained based on a standard man plan, there is an increased risk of injury which may lead to session failure or quitting the training.

The training mesocycle presented in Table 7.2 is suitable for a beginner trainee; there are also two other types of mesocycle: development, more intense than the first and, realization, which usually includes a challenge or a competition. In the realization mesocycle, most of the session loads are light, while the competition days have the biggest load. The mesocycle sequences are the same for men. The sequences are as follows for a beginner trainee: beginning-development-realization and then they follow a development-realization loop: B-D-R-(D-R).

The importance of sex-specific tutoring is demonstrated through a simulation with hundreds of women having randomized menstrual days. A comparison between standard training and optimal woman training has been conducted, and the accumulated loss was computed. Women and men are virtually trained for 240 days, based on a three sessions per week plan. The optimal sessions load for a beginner woman and for a beginner man starting on the same day can be seen in Fig. 7.6a, b. The 1–5 values on the axis show the session load type: 1—light, 2—medium, 3—heavy, 4—very heavy, and 5—challenge session.

The bars on Fig. 7.6c show the difference between men and women training; for example, the first blue bar shows a man session load equal to 2, and a woman session load equal to 1. The sessions with blue bars show these have a risk of injury and potential session failure. For the presented simulation, the accumulated risk of injury loss was -49. Sessions with no associated color (value zero) show that the proposed session is optimal—matching men and women optimal load. The overall count of loss in sessions (values different than zero) is 82, whereas the cumulated

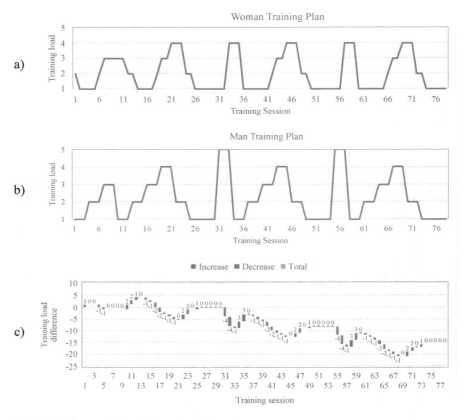

Fig. 7.6 Differences in simulations for training plan

differences are −16. This shows that man training planification applied to women is suboptimal and can lead to injuries; thus, sexual dimorphism training sequences should be considered.

7.5 Conclusions and Future Work

Understanding the mechanics of motor skill acquisition is critical for designing an intelligent tutoring system for psychomotor skills development. Such a smart learning environment should also consider the sexual dimorphism when training for motor skill acquisition, the sexual dimorphism being demonstrated by several morphological and physiological differences. Current work presents updates for the *Selfit* system used for psychomotor skills development by integrating the sexual dimorphism in the ITS architecture: student, domain, and tutoring modules.

The updated system proposes a monitoring module which is using computer vision technology for assessing the women ACL risk of injury. It also presents an update for the tutoring module, which uses men and women mesocycle calendars, based on the menstrual date and periodicity. Standard training plans use the same training rules for both men and women, while the sport science has proven the contrary, with risks of suboptimal training, injury, or training failure. As the simulation shows promising insights, future work should follow up the contribution to the smart learning environments used for psychomotor skills development. Future experiments may focus on mass user testing on the updated *Selfit* system and data collection to assess the efficiency of the training personalization and accounting for sexual dimorphism. Also, the monitoring of the morphological and physiological trainee body parameters, while performing personalized training, may provide valuable insights, which can open new research directions for the sport science.

References

1. Kirchengast, S.: Gender differences in body composition from childhood to old age: an evolutionary point of view. J. Life Sci. **2**(1), 1–10 (2010)
2. Wells, J.C.K.: Sexual dimorphism in body composition. Best Pract. Res. Clin. Endocrinol. Metab. **21**, 415–430 (2007)
3. Neagu, L.-M., Rigaud, E., Guarnieri, V., Travadel, S., Dascalu, M.: Selfit—an intelligent tutoring system for psychomotor development. In: International Conference on Intelligent Tutoring Systems (2021)
4. Goldberg, B., Amburn, C., Ragusa, C., Chen, D.-W.: Modeling expert behavior in support of an adaptive psychomotor training environment: a marksmanship use case. Int. J. Artif. Intell. Educ. **28**(2), 194–224 (2018)
5. Brawner, K., Hoffman, M., Nye, B.: Architecture and ontology in the generalized intelligent framework for tutoring: 2018 Update. In: 7th Generalized Intelligent Framework for Tutoring (GIFT) Users Symposium, p. 11. US Army Combat Capabilities Development Command–Soldier Center (2019)
6. Neagu, L.-M., Rigaud, E., Travadel, S., Dascalu, M., Rughinis, R.-V.: Intelligent tutoring systems for psychomotor training—a systematic literature review. In: 16h International Conference on Intelligent Tutoring Systems (ITS 2020). Springer, Online (2020)
7. Joyce, D., Lewindon, D.: Sports Injury Prevention and Rehabilitation. Integrating Medicine and Science for Performance Solutions. Taylor and Francis, UK (2016)
8. Shephard, R.J.: Exercise and training in women, Part I: influence of gender on exercise and training responses. Can. J. Appl. Physiol. **25**(1), 19–34 (2000)
9. Baron-Cohen, S., Knickmeyer, R.C., Belmonte, M.K.: Sex differences in the brain: implications for explaining autism. Science **310**(5749), 819–823 (2005)
10. Liutsko, L., Muiños, R., Tous Ral, J.M., Contreras, M.J.: Fine motor precision tasks: sex differences in performance with and without visual guidance across different age groups. Behav. Sci. **10**(1), 36 (2020)
11. Li, R.: Why women see differently from the way men see? A review of sex differences in cognition and sports. J. Sport Health Sci. **3**(3), 155–162 (2014)
12. Chidi-Ogbolu, N., Baar, K.: Effect of estrogen on musculoskeletal performance and injury risk. Front. Physiol. **9** (2019)
13. Pitchers, G., Elliott-Sale, K.: Considerations for coaches training female athletes. Professional Strength & Conditioning, Training Female Athletes (2019)

14. Somerson, J.S., Isby, I.J., Hagen, M.S., Kweon, C.Y., Gee, A.O.: The menstrual cycle may affect anterior knee laxity and the rate of Anterior Cruciate Liga-ment Rupture: a systematic review and meta-analysis. JBJS Rev. **7**(9), e2 (2019)
15. Neagu, L.-M., Guarnieri, V., Rigaud, E., Travadel, S., Dascalu, M., Rughinis, R.-V.: An ontology for motor skill acquisition designed for GIFT. In: Proceedings of the 8th Annual Generalized Intelligent Framework for Tutoring (GIFT) Users Symposium (GIFTSym8) (2020)
16. Cao, Z.H., Hidalgo, G., Simon, T., Wei, S.-E., Sheikh, Y.: OpenPose: Realtime Multi-Person 3D Pose Estimation Using Part Affinity Fields. arXiv preprint arXiv:1812.08008 (2018)
17. Simon, T., Joo, H., Matthews, I., Sheikh, Y.: Hand keypoint detection in single images using multiview bootstrapping. In: Proceedings of the IEEE Conference on Computer Vision and Pattern Recognition (CVPR), pp. 1145–1153 (2017)
18. Ota, M., Tateuchi, H., Hashiguchi, T., Kato, T., Ogino, Y., Yamagata, M., Ischihashi, N.: Verification of reliability and validity of motion analysis system during bilateral squat using human pose tracking algorithm. Gait Posture **80**, 62–67 (2020)

Chapter 8
Romanian Syllabification Using Deep Neural Networks

Dragos-Georgian Corlatescu, Stefan Ruseti, and Mihai Dascalu

Abstract Syllabification may be considered trivial for humans, but it can prove to be a challenging task in terms of automated text analysis. In this study, we explore three approaches to syllabify words in Romanian using state-of-the-art deep learning architectures in sequence prediction, namely BiLSTM, CNN, and transformer. In contrast to previous approaches, our models take into account the part of speech of the word, which in return can weigh heavily in situations where words have the same written form, but different syllabification. Our best model obtains an accuracy of approximately 98% using a conditional random field on top of the BiLSTM architecture, surpassing all previous state-of-the-art models. Our model represents a building block for multiple smart learning ecosystems, ranging from better hyphenation software for text evaluation, to text-to-speech and speech-to-text frameworks employed in intelligent houses or personal assistants.

Keywords Syllabification for Romanian · BiLSTM · CNN · Transformer-based architecture · CRF

8.1 Introduction

Syllabification is the process of splitting a word into syllables that are generally marked by a hyphen. The division points are language specific and, in general, can be found by humans without consulting a dictionary. However, there are cases where

D.-G. Corlatescu · S. Ruseti · M. Dascalu (✉)
Computer Science Department, University Politehnica of Bucharest, 313 Splaiul Independentei, 060042 Bucharest, Romania
e-mail: mihai.dascalu@upb.ro

D.-G. Corlatescu
e-mail: dragos.corlatescu@upb.ro

S. Ruseti
e-mail: stefan.ruseti@upb.ro

M. Dascalu
Academy of Romanian Scientists, Str. Ilfov, Nr. 3, 050044 Bucharest, Romania

Ó. Mealha et al. (eds.), *Ludic, Co-design and Tools Supporting Smart Learning Ecosystems and Smart Education*, Smart Innovation, Systems and Technologies 249,
https://doi.org/10.1007/978-981-16-3930-2_8

93

correct syllabification is not easily determined—for example, there are words that sound equally right in two or more hyphenated forms. Syllabification enhances other research topics, such as rhythm analysis or text-to-speech conversion. For example, the process can be further integrated in an artificial intelligence assistant communicating with the user via spoken language in various environments, including a smart home. Moreover, intelligent tutoring systems can integrate such functionalities to help children learn to syllabify by proving automated syllabification materials, tests, and evaluations.

The syllabification problem involves discovering generalized patterns in words. Language experts can enumerate all the possible word syllabifications, but this approach is time consuming for the entire dictionary of a language. As such, an automated model with high accuracies represents a viable alternative, and two emerging approaches exist as follows: (a) enforcing a predefined set of rules applied like a finite state machine and (b) statistical methods.

The first approach may work for some languages, but that is not the case for Romanian as there is no generalizable scripts for splitting Romanian words into syllables. There are specific rules that can be applied to support the syllabification process, such as (a) a syllable must have one and only one vowel and (b) consonants and/or semivowels can appear alongside the vowel. DOOM 2 (The Orthographic, Orthoepic and Morphological Dictionary of the Romanian Language) [1] also specifies six rules that represent decisions for specific cases. However, two rules are circular and cannot be used in practice since they are relying on the concepts of diphthong and triphthong to split the words; nevertheless, knowing the syllables is required to detect either of two previous concepts.

The second approach involves statistics for a model to learn how to syllabify by providing multiple examples from which it can generalize. Pure statistical or artificial intelligence approaches, such as neural networks, can be considered; this study abides to this approach and our model for syllabification in Romanian integrates state-of-the-art deep learning architectures in sequence prediction, while also accounting for the part of speech of each word.

The paper continues with the presentation of previously applied methods for the syllabification task. The third section describes the used corpora, alongside the method employed for building the models. Afterward, the paper continues with the conducted experiments, followed by an analysis of the results. Finally, conclusions are presented together with future work ideas.

8.2   Related Work

There are three main approaches to perform word syllabification: (a) dictionary-based, (b) rule-based, and (c) statistical models. The dictionary-based method keeps in the computer's memory a predefined mapping between words and their correct syllabification. If this method is used, we can be certain that the obtained syllables are reliable, except in a small number of cases where human errors may be encountered.

On the downside, this method requires a lot of manual work, and it can become cumbersome with the expansion of the dictionary, as well as the syllabification of proper nouns. However, the method is straightforward, and there is no need for additional research.

In general, languages have different sets of norms when speaking about syllabification; as such, rule-based systems may approach the problem with different hypothesis. Nevertheless, understanding the methods applied on other languages can provide valuable insights.

The work of Mañas [2] is representative for the Spanish language. Their study presents requirements for syllabifying in Spanish, while relating to letters or group of letters. Additionally, an algorithm is implemented in the C programming language that follows the steps and the rules provided by the Royal Academy of Spanish Language. As such, Spanish is a fortunate case that does not require any other advanced methods for automated word syllabification.

Compound words are frequent in German, making the dictionary-based approach almost impossible to implement because of the numerous words formed via composition. Kodydek [3] focused on splitting compound words in "atoms" that can be considered similar to morphemes. Their method improves a rule-based model by marking syllabification points before and/or after an atom.

The development of a finite state machine based on syllabification rules was explored for Dutch [4]. The rules were represented as regular expressions that can be applied on the word and the machine-generated syllabification points based on the matches. Transformation-based learning [5] was also applied to improve the results to an accuracy score above 99%.

The interest in word syllabification for English was enlarged by the text editing software companies for which word splitting at the end of a row is important, due to its implications on the overall aspect of the documents. Liang [6] focused on implementing a more robust syllabification method for TEX, a system for technical writing. But rather than focusing just on the syllabification problem, there was another aspect that could not be ignored at that time in history: limited physical memory space. The method tries to minimize the stored information, while increasing performance using patterns (rules). In the same line or research, Rosenbaum [7] proposed a digital reference hyphenation matrix to limit the used memory.

The aspect regarding hyphenation of printed documents was explored in multiple patents. Ranking hyphenations was studied by Hersey et al. [8], where choosing the right hyphen point was done automatically with the focus on the visual aspect of the document. Similar, Carlgren et al. [9] combined three methods (dictionary-based, ruled-based, and algorithm-based) to determine the correct hyphenation and then applied a correct right margin justification.

Artificial intelligence methods, especially neural networks, are preponderantly used nowadays. Alonichau et al. [10] described in their patent a general architecture to train a multilingual system for syllabification. A concrete example was implemented for both English and Dutch language [11] using conditional random fields (CRF) [12] that outperformed the rule-based systems at the time of the publishing.

The best results for English syllabification were obtained by Bartlett et al. [13] who used a support vector machine (SVM) alongside a hidden Markov model to predict syllables. The starting point in the representation was the phoneme in which each letter was tagged by its importance: The vowel was the nucleus, while the optional preceding letters were the onset, and the optional following letters were the coda. Barlett et al. also obtained exceptional results on German and Dutch languages with an accuracy over 99%.

Barbu [14] created a corpus for Romanian called RoSyllabiDict that represents the foundations of our experiments. An actual implementation of a syllabification model was done by Dinu et al. [15] using support vector machines and conditional random fields that obtained a 95.25% word accuracy. Also, Boroş [16] applied syllabification with the margin infused relaxed algorithm (MIRA), where the numbered onset-nucleus-coda (ONC) was the chosen method for representing the syllables; Boroş obtained an accuracy of 99.01% on OOV words. Boros et al. [17] used the same representation to obtain the state-of-the-art for Romanian syllabification task using an improved ID3 algorithm—99.17%.

8.3 Method

8.3.1 Corpora

We first considered the corpus of Barbu [14]. However, the corpus contained many errors that required manual corrections, and there were still errors unsolved even after multiple iterations of cleaning the corpus. These errors affected the performance of our models, and due to this reason, we switched to the Romanian Academy Explanatory Dictionary that contained approximately 600 K manually verified, syllabified words. The words were grouped based on their lemma, and when the lemma was identical, their corresponding part of speech (POS) was the second differentiator. Starting from the base dictionary form, all the derived forms were syllabified. If a word had multiple parts of speech, then different entries were present in the corpus.

The corpus was split into train, development, and test subsets using two approaches. The first one ensured that two derived forms from the same lemma were always in the same dataset. The only exception was when a word (lemma) had multiple POS tags—in that case, the words with the same POS were always together and in a different subset than the syllabifications coming from the other parts of speech. The second split was a random split that did not have additional rules. The split was 60% (train)/20% (dev)/20% (test) in both cases. Note that, at least in theory, the first split is harder, and at the same time, more relevant for machine learning purposes as it prevents overfitting by grouping words with similar syllabifications in the same subset.

8.3.2   Neural Architecture

The problem of syllabification can be described as taking a sequence of letters as input and returning another sequence, with the hyphens in the correct place, as output. With this in mind, our first approach considered a sequence-to-sequence neural network [18], where the model was fed each character of the word at a time and return the hyphenated version, character by character. However, the obtained results using this architecture were not great, which is not surprising considering that the model was too complex for the given task. A sequence-to-sequence model needs to learn to repeat all characters in the input word besides learning when to output a hyphen.

Therefore, we modeled the problem as a binary classification for each of the characters in the input word. An output of 1 for a character means that it represents the end of a syllable, and as such, a hyphen needs to be added afterward (except for the last letter); otherwise, the model would predict a 0 for all the other letters. In this problem definition, the model needs only to focus on the hyphens, without the need to learn how to generate all the letters in the word. In addition, the transformation into a classification problem opens the possibility of using a larger spectrum of architectures, besides sequence to sequence.

The following steps were applied to format the input training data and make it suitable for a neural network:

1. Special characters, namely "\t" and "\n," were added to all the words (syllabified or not) to symbolize the beginning and, respectively, the end of the sequence.
2. The maximum sequence length was computed so that the input and output arrays have the same size.
3. All possible characters were mapped to an integer number and stored in a dictionary.
4. All possible POS tags were mapped to an integer number and stored in a dictionary.
5. Input words were formatted to represent a sequence given by the mapping at step 3.
6. POS tags were replaced by a number from the dictionary computed at step 4.

The output data was formatted to a 0 and 1 sequence with the size of "maximum sequence length" calculated at step 2. A padding with 0 was applied to ensure consistency between all output words.

All neural models configurations started with a character embedding layer of size 16 in the initial experiments. The next layer was an encoding layer that computed contextualized character representations, each of them followed by the same dense layer of size 8 with RELU as the activation function. Finally, a dense layer with a sigmoid activation function was used for the binary classification of each character. The optimizer chosen was Adam with a learning rate of 1e−4, while binary cross-entropy was used as the loss function.

Three different architectures for computing contextualized character embeddings were evaluated as follows: bidirectional long-short term memory (BiLSTM) [19, 20], convolutional neural networks (CNN) [21], and transformer [22]. For the BiLSTM, a cell size of 16 was used, the CNN used 16 kernels of length 6, while the transformer layer had 8 attentions heads, and the hidden layer in the feed-forward network within the transformer had 32 units. A positional embedding was also included as input, because the transformer does not keep track of the position of characters by default.

Even though the results were promising, as presented in next section, there was a major problem: Two words with the same written form were syllabified exactly the same, even though they have different meanings or parts of speech—for example, "haină" syllabified as "ha-i-nă" (eng., "wicked"), *versus* "haină" syllabified as "hai-nă" (eng., "coat"). Therefore, we decided to integrate the word part of speech as an input in the neural network. Thus, each previous model presented changed from a sequential model to a functional model that allowed two inputs: the characters sequence and the codified POS. The part of speech was fed into another embeddings layer that differs in dimension from 2 (i.e., when only the general POS was given, for example, "NOUN") to 16 (i.e., when the specific POS was considered, for example, "Ncms-n").

The final (and best) model involved an additional conditional random fields (CRF) layer placed on top of the last dense layer with the RELU activation function (see architecture in Fig. 8.1). The model chosen for this experiment was the BiLSTM due to its higher performance in previous stages. The Viterbi sequence obtained from the

Fig. 8.1 Neural architecture including POS and character embeddings, BiLSTM, and CRF (POS + BiLSTM + CRF)

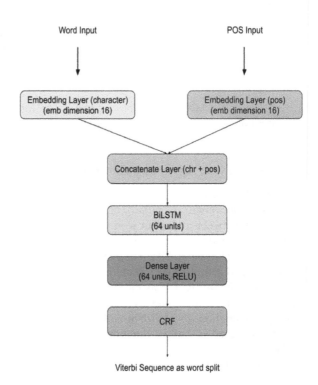

CRF layer was integrated in a custom metric that returned a score of 1 if the whole syllabification was correct or 0 otherwise.

8.4 Results and Discussion

Most experiments were conducted on the lemma split corpus because it reduces the chances for model overfitting. Table 8.1 presents the accuracies of our models which are incremental in terms of configuration, on each split of the dataset, namely train, dev, and test.

The first three models were the simpler ones that did not have additional information, such as the part of speech tags of the word. The results showed good prediction scores, with the BiLSTM model performing slightly better than the other two alternatives. The convolutional neural network was performing almost as good as the BiLSTM, while being the fastest to train. These results argue that in most cases, looking only at 2–3 characters before and after the target letter is enough to predict the correct split. The transformer model was still not capable of performing as well as the other two models, despite adding positional encoding as input. This was expected, since long-distance dependencies are not important for this task, while the exact order of letters is crucial.

The next six results are obtained when using word POS tags as input. There are two approaches for the part of speech to be concatenated to the network: (a) combine POS embeddings with the character embeddings which are then fed into the contextualized character embedding layer (i.e., POS before BiLSTM/CNN/Transformer) or (b) the result of the contextualized embedding layer is combined with the POS embedding at the end (POS after BiLSTM/CNN/Transformer). Adding POS information leads

Table 8.1 Word level accuracies on the lemma split corpus

Model	Train acc. (%)	Dev acc. (%)	Test acc. (%)
BiLSTM	97.32	96.13	95.98
CNN	96.94	95.70	95.32
Transformer	96.29	96.11	95.76
POS + BiLSTM (POS before BiLSTM)	97.73	96.49	96.01
BiLSTM + POS (POS after BiLSTM)	97.76	96.42	96.13
POS + CNN (POS before CNN)	97.30	95.82	95.99
CNN + POS (POS after CNN)	97.29	95.74	95.77
POS + Transformer (POS before Transformer)	96.48	96.15	95.64
Transformer + POS (POS after Transformer)	96.46	96.11	95.59
POS + BiLSTM + CRF	**99.95**	**98.32**	**97.55**
MLPLA (ID3) [17]	96.97	97.30	97.07

The bold values denote the best performing model

Table 8.2 Word level
accuracies on the random
split corpus

Model name	Train acc. (%)	Dev acc. (%)	Test acc. (%)
POS + BiLSTM + CRF	**99.92**	**99.67**	**99.70**
MLPLA (ID3)	97.09	97.06	97.02

The bold values denote the best performing model

to an increase in accuracy of roughly 0.4% in all configurations. The differences
between the positions where the POS information was added into the network were
not that significant, with slightly higher scores when combining the POS embedding
with the character embedding before the contextualized character embeddings layer.

Our last configuration (POS + BiLSTM + CRF) considers the best previous setup
(POS + BiLSTM) on top of which a CRF layer was added (see Fig. 8.1). Empirical
adjustments in terms of layers sizing and learning rate (1e−3) were made.

The last line of Table 8.1 represents the scores obtained using the model by that
currently holds the state-of-the-art results [17]. The model is freely available online
at: https://slp.racai.ro/. Unfortunately, the authors reported their results on a different
corpus to which we do not have access; as such, the provided pre-trained ID3 model
was applied on all entries from our three partitions of the dataset. Our best performing
model surpasses their performance.

In order to further argue the benefits of our method, we consider also the random
split into train, dev, and test with the same 60–20–20 ratios. Table 8.2 shows the
comparison between the previous state-of-the-art model and our best configuration that now exceeds 99.60% on all partitions. Even when considering the original
reported results by Boros et al. [17] of 99.21% (train), 99.07% (dev), and 99.16%
(test), our model achieves higher scores. Additionally, these result show that the
lemma split corpus is indeed harder and, at the same time, more relevant. Nonethe-
less, the models presented in this paper obtained good results in both cases, denoting
that they generalize properly the task at hand.

8.5 Conclusions and Future Work

Multiple neural models were developed for the Romanian syllabification task
grounded in the idea of sequence to binary classification. The main three layers used
for building the models were the BiLSTM, CNN, and the transformer. Part of speech
tags were also considered besides the words themselves. The approach within our
experiments was incremental: We started from the models that only used the words
as input and then added the part of speech as an additional feature, followed by a
final CRF layer. The POS tags of words increased the overall accuracies by approximately 0.4%. The model that obtained the highest results was the POS + BiLSTM +
CRF configuration, which surpasses the previous state-of-the-art ID3 model: 98.35%
versus 97.30% on dev, and 97.55% versus 97.07% on test. These experiments were

run on a train, dev, and test split that took into consideration the lemmas of the words, making it harder for a model to overfit. Another run was performed on a randomly split corpus and, as expected, accuracies were higher (more than 99.60% on each split).

This research introduces a new model that obtains state-of-the-art results on the syllabification task for Romanian, while also accounting for the POS tag of each word. This approach is even more relevant because words generally appear in specific contexts, and automated POS tag models already have achieved high accuracies (e.g., 96% for Romanian [23]).

In terms of future work, we want to integrate our model in related tasks, such as determining the rhythm of the texts, measuring the complexity of words, or performing speech to text synthesis. These natural language processing tasks represent the steppingstones for more complex projects in the smart learning ecosystem field, such as building a conversational agent for a smart home environment or developing a platform to support children in the process of learning how to properly syllabify. In addition, we want to extend our approach to other languages, while also accounting for emerging patterns of potential similarities or differences between languages.

Acknowledgements This work was funded by a grant of the Romanian National Authority for Scientific Research and Innovation, CNCS—UEFISCDI, project number TE 70 PN-III-P1-1.1-TE-2019-2209, ATES—"Automated Text Evaluation and Simplification."

References

1. Vintila-Radulescu, I. (ed.): DOOM2. Dicționarul ortografic, ortoepic și morfologic al limbii române. Univers Enciclopedic, Bucharest, Romania (2005)
2. Mañas, J.A.: Word division in Spanish. Commun. ACM **30**(7), 612–616 (1987)
3. Kodydek, G.: A word analysis system for German hyphenation, full text search, and spell checking, with regard to the latest reform of German orthography. In: International Workshop on Text, Speech and Dialogue, pp. 39–44. Springer, Brno, Czech Republic (2000)
4. Bouma, G.: Finite state methods for hyphenation. Nat. Lang. Eng. **9**(1), 5–20 (2003)
5. Brill, E.: Transformation-based error-driven learning and natural language processing: a case study in part-of-speech tagging. Comput. Linguist. **21**(4), 543–565 (1995)
6. Liang, F.M.: Word hyphenation by computer. PhD, Department of Computer Science, Stanford University, Stanford, CA, USA (1983)
7. Rosenbaum, W.S.: Digital reference hyphenation matrix apparatus for automatically forming hyphenated words. In: USPTO (ed.). Google Patents, United States (1977)
8. Hersey, I.L., Stephens, R.L., Zamora, A.: Computer method for ranked hyphenation of multilingual text. In: USPTO (ed.). Google Patents, United States (1994)
9. Carlgren, R.G., Reed, M.A., Rosenbaum, W.S.: Mixed mode enhanced resolution hyphenation function for a text processing system. In: USPTO (ed.). Google Patents, United States (1986)
10. Alonichau, S., Shahani, R., Powell, K.: Multi-lingual word hyphenation using inductive machine learning on training data. In: USPTO (ed.). Google Patents, United States (2015)
11. Trogkanis, N., Elkan, C.: Conditional random fields for word hyphenation. In: Proceedings of the 48th Annual Meeting of the Association for Computational Linguistics, pp. 366–374. The Association for Computer Linguistics, Uppsala, Sweden (2010)

12. Lafferty, J., McCallum, A., Pereira, F.C.: Conditional random fields: probabilistic models for segmenting and labeling sequence data. In: 18th International Conference on Machine Learning, pp. 282–289. Morgan Kaufmann, Williamstown, MA, USA (2001)
13. Bartlett, S., Kondrak, G., Cherry, C.: Automatic syllabification with structured SVMs for letter-to-phoneme conversion. In: Proceedings of ACL-08: HLT, pp. 568–576. The Association for Computer Linguistics, Columbus, OH, USA (2008)
14. Barbu, A.-M.: Romanian lexical data bases: inflected and syllabic forms dictionaries. In: International Conference on Language Resources and Evaluation. European Language Resources Association, Marrakech, Morocco (2008)
15. Dinu, L.P., Niculae, V., Sulea, O.-M.: Romanian syllabication using machine learning. In: International Conference on Text, Speech and Dialogue, vol. 8082, pp. 450–456. Springer, Pilsen, Czech Republic (2013)
16. Boroş, T.: A unified lexical processing framework based on the Margin Infused Relaxed Algorithm. A case study on the Romanian Language. In: Proceedings of the International Conference Recent Advances in Natural Language Processing RANLP 2013, pp. 91–97. RANLP 2013 Organising Committee/ACL, Hissar, Bulgaria (2013)
17. Boros, T., Dumitrescu, S.D., Pipa, S.: Fast and accurate decision trees for natural language processing tasks. In: RANLP, pp. 103–110. INCOMA Ltd., Varna, Bulgaria (2017)
18. Sutskever, I., Vinyals, O., Le, Q.V.: Sequence to sequence learning with neural networks. In: Advances in Neural Information Processing Systems, pp. 3104–3112. Curran Associates, Inc., Montreal, Quebec, Canada (2014)
19. Schuster, M., Paliwal, K.K.: Bidirectional recurrent neural networks. IEEE Trans. Signal Process. **45**(11), 2673–2681 (1997)
20. Hochreiter, S., Schmidhuber, J.: Long short-term memory. Neural Comput. **9**(8), 1735–1780 (1997)
21. Krizhevsky, A., Sutskever, I., Hinton, G.E.: Imagenet classification with deep convolutional neural networks. Adv. Neural. Inf. Process. Syst. **25**, 1097–1105 (2012)
22. Vaswani, A., Shazeer, N., Parmar, N., Uszkoreit, J., Jones, L., Gomez, A.N., Kaiser, Ł., Polosukhin, I.: Attention is all you need. In: Advances in Neural Information Processing Systems, pp. 5998–6008. Curran Associates, Inc., Long Beach, CA, USA (2017)
23. Honnibal, M., Montani, I.: spaCy 2: natural language understanding with bloom embeddings. Convolutional Neural Networks and Incremental Parsing **7**(1) (2017)

Part III
Observing and Studying Learning Ecosystems

Chapter 9
A Year After the Outbreak of COVID-19: How Has Evolved the Students' Perception on the Online Learning?

Carlo Giovannella

Abstract During the last year students have been exposed to a continuous evolution of the learning processes (fully online, blended, parallel blended, rotation of the face-to-face activities, etc.) and, often, to a lot of different technologies, with only one common denominator: the carrying out of the educational activities. Despite the peculiarity of the pandemic situation, one year is a time long enough to analyze the evolution of the students' perceptions and understand if could be put in relation with a possible start-up of processes of technology innovation and/or acceptance. In this article, we present the outcomes of an investigation conducted by means of an online survey at almost one year distance from the first one that involved, like this time, the students of the bachelor course in Educational Science of the University of Tor Vergata: this year 159 students (98% women). We have been able to explore the relevance of many categories of factors: intrinsic to the characteristics of the learning ecosystems, personal, contextual, related to the educational activities, perceived values of the technology-based processes. It appears that the main causal chain moves from the characteristics of the learning ecosystem, toward the impressions about the learning activities and their effectiveness to, finally, the future intentions and expectations. Not very relevant appear to be individual and contextual factors. The overall impressions of the students about the online activities seem to be positive but personal setting (equipment and connectivity) appear to be more effective than that provided by the university, possibly to underline criticalities met with the delivering of parallel blended activities. In addition, students deem very importantly the collaboration/interaction among peers and with the teachers and, as well, collaborative and design activities, while, on the contrary, most of the implemented learning activities seem to have a transmissive nature.

Keywords Online learning · Blended learning · Parallel blended learning · Perception about technologies · University students · Technology acceptance · Technological innovation · Technology-based learning · Smart learning

C. Giovannella (✉)
University of Rome Tor Vergata—Dip. SPSF, Rome, Italy
e-mail: gvncrl00@uniroma2.it

ASLERD, Rome, Italy

© The Author(s), under exclusive license to Springer Nature Singapore Pte Ltd. 2022
Ó. Mealha et al. (eds.), *Ludic, Co-design and Tools Supporting Smart Learning Ecosystems and Smart Education*, Smart Innovation, Systems and Technologies 249,
https://doi.org/10.1007/978-981-16-3930-2_9

ecosystems · Covid-19 pandemic · Future perspectives · Descriptive analysis ·
Causal discovery · MAETI

9.1 Introduction

One year after the outbreak of the pandemic and the sudden transfer of the educational activities from physical spaces to virtual environments [1], we have witnessed
continuous adjustments in the delivery of educational processes [2, 3]. Politicians,
for both economic and social reasons, attempted and are still attempting to restore
face-to-face learning activities using different modalities, depending on the different
level of the curricula (K12, high-school and university). In the case of the universities,
the decision makers attempted to offer face-to-face activities to the first-year students
by limiting the number of attendants and rotating them. In parallel teachers started
to test the delivery of what we could define "parallel blended modality" in which
the activities are attended simultaneously by students in the classroom and those
connected from home. Any sort of attempt to resume face-to-face activities has been
characterized by "stop and go" induced by the evolution of the pandemic and the use
of temporary zonal lockdowns. Regardless of the heterogeneity of the didactic practices that have been put in place, most of them have been continuously supported by
the use of technologies, that during this last year have marked the succession of days
for both students and teachers. Because of this, it is very important to understand if
and how the perception of the actors involved in the learning processes has evolved
in one year time. A good starting point is represented by one of the first case studies
that have been conducted in Italy—one month after the pandemic outbreak [4]—that
involved the students of the Bachelor of Science of Education of the University of
Rome Tor Vergata. Since that time the investigation tools and the methodologies
of analysis that we adopted have been improved [5–7] to identify also the causal
relations among factors that could support or prevent the evolution of the innovation
process triggered by the outbreak of the pandemic.

As well described by the literature [8], in fact, an innovation process that involves
the adoption of "new" technologies is a long process that must go through a certain
number of phases and involves the concomitant intervention of numerous factors.
At present, these factors have not been completely and clearly identified both in
their nature and in their relations. Numerous models have been proposed [9] and
several studies have been conducted to demonstrate the relevance of the aforementioned factors. All such studies, however, have been performed always in situations in
which technologies have been introduced in a controlled and progressive manner with
the agreement of the involved actors. In the case of the learning processes carried on
during the pandemic, however, the situation has been quite different: the involvement
of individuals was sudden and forced and, moreover, there was no precise definition
of the technologies that could be used or of the didactic activities to be implemented
with the support of such technologies. We were faced with a completely anomalous

situation with respect to the usual processes of technology innovation and/or adoption. In such a situation it is not at all obvious that the factors related to the innovation and acceptance models developed previously could be applied straightforwardly.

Accordingly, the investigation described in this article has three integrated goals:

(a) the monitoring of the evolution of the students' perceptions, keeping in mind that one year ago [4]—since we needed to "photograph" the sudden transition from the face-to-face to online didactic activities—we had not enough time to design a survey aimed at identifying which could be the most relevant factors;

(b) an exploration about the relevance assumed by the factors that have been taken into consideration by previous well-known models of technological innovation and acceptance (see next section);

(c) the verification of MAETI (Model for Attitude to get Engaged in Technological Innovation) that we have derived in Ref. [7].

9.2 Experimental Setting

9.2.1 Questionnaire

To design the questionnaire used in the present study we started by taking into consideration the various categories of factors that have been used in the past to investigate and describe the processes of *technology innovation and acceptance*. A very nice and synthetic overview of the main models available in the literature has been provided in a recent article by Leoste et al. [9] and includes: the *Technology Acceptance Model* (TAM) [10], the *Technology-Organization-Environment* (TOE) framework [11], the *Unified Theory of Acceptance and Use of Technology* (UTAUT) [12] and the *Diffusion of Innovation Theory* (DOI) [8]. Here, due to the limitation of space, we restrict ourselves only to briefly underline the peculiarities that characterizes the pandemic situation and that can be ascribed to the sudden need to swap to technology enhanced and on-line learning.

Innovation process: during the pandemic there has been no time to fully develop an *awareness* phase; all actors (students, teachers, parents, etc.) have been forced to *get aware* and, at the same time, *use* technologies as active protagonists of the educational processes. Technology *acceptance* has been forced and, at present, it is not clear to which extent it could be ever transformed in *adoption*. This is one of the open questions that need to be investigated through the monitoring of actors' future intentions, keeping in mind that most likely any intention of adoption will be probably counteracted by multiple forces that are pushing to get back to a "new normality" as soon as possible.

Relevant factors: future intentions may be influenced by multiple categories of factors (see Table 9.1) related to the *context* (i.e. the *learning ecosystem*), to the *individuals*, to the *educational process* and finally to the *perceived values*. Here we

Table 9.1 List of the main factors considered in the present case study, organized by domains

Factor domain	Factors considered in our investigation	Correlated models
Learning ecosystem: technological setting	University connectivity (UC); technological adequacy of university (TAU; TAOE in Ref. [5])	DOI and UTAUT
Learning ecosystem: organizational factors	University reaction capacity (URC); emergency management (EM; UR in Ref. [5]); operational assistance (OA)	TAM, DOIT, UTAUT and TOE
Learning ecosystem: competences	Teachers' technological preparedness (TTP; TTR in Ref. [5]); teachers' technological preparedness (TTP; TTR in Ref. [5])	
Personal factors	Personal connectivity (PC); individual technological adequacy (ITA); individual technological preparedness (ITP); individual innovation propensity (IIP); self-regulation capability (SRC)	Partially considered by DOI and UTAUT
Contextual factors: wider environment	Political action (PA); public opinion (SI); influence of the prevalent opinion in the context (IPOC)	DOI and TOE
Contextual factors: individual level	Influence of familiar environment (IFE); psychological problems (PSP)	
Educational activities/processes	Didactic activity integration (DAI); reproducibility of classroom dynamics (RCD); collaborative activities (CA); collaborative activities quality (CAQ); collaboration with peers (CP); communication with teachers (CWT); communication with peers (CWP); relationships with peers (RWP); relationships with teachers (RWT); videoconf usage for lesson (VUL) videoconf usage for discussion (VUD); forum usage (FU); design activities (DA)	Partially considered by the knowledge appropriation model (KAM) [13]

(continued)

Table 9.1 (continued)

Factor domain	Factors considered in our investigation	Correlated models
Perceived values of technology-based processes	Easyness of use of the integrated didactic (EUID; EUOL in Ref. [5]); usefulness of the integrated didactic (UID; UOL in Ref. [5]); usefulness of technologies (UOT)	TAM, UTAUT, DOI and TOE
Perceived efficacy	Efficacy of use of the integrated didactic (EID); change in efficacy of the evaluation (CEE)	TAM, DOI and TOE
Perceived changes (individual and process levels)	Motivation toward didactic activities (MDA); personal time management capacity (PTMC; TTMC in Ref. [5]); improvement in the feeling toward technologies (IFT); improvement in technological skills (ITS); change in didactic activities (CDA); change in evaluation modality (CEM); workload increase (WI)	
Outcomes: learning ecosystems	University innovation propensity (UIP); degree of university e-maturity (UeM)	
Outcomes: learning processes	Extent to which university should rely on integrated didactic (URID); sustainability of online education (SOE); usefulness of education on integrated didactic (UEID; REDP in Ref. [5])	
Outcomes: individual intentions	Intention to use integrated didactic (IEID); intention to work in smart working (IWSW); percentage of use integrated didactic (PUID)	

In the third column are shown the potential relations with technology innovation and acceptance models

wish to underline that in the case of the context considered here, the *technological factors* have to be associated with the overall technological infrastructure and apparats made available by the learning ecosystem, and not with a single well-defined technology. To describe the setting, moreover, technological factors should be integrated with *organizational factors* and with the available *competences*. In addition, we would like to note that in our case study we should consider both extrinsic *social factors* and intrinsic ones; these latter cannot be considered separately from the learning processes that are likely to be shaped by social interaction, scaffolding and collaboration. Moreover, due to the wide diversification of the learning processes that have been implemented during the pandemic, it is important to investigate how such variability could have influenced the causal relations among factors. Finally, we want to remark that any factor that one could consider may assume a different relevance depending on the category of the actors (students, teachers, parents, etc.) involved in the survey. In this work, we consider the point of view of university students. Keeping all the above remarks in mind, together with the outcomes of previous research [5–7], we have designed a survey to investigate the relevance of the factors listed in Table 9.1.

The questionnaire is composed by of five sections and comprises 82 items. The first section comprises 2 socio-biographical background items (gender, age). The second one explores the setting and presents 25 items (19 questions requiring a multiple choice or numerical answer and 6 open questions or requests for explanatory comments). The third section is dedicated to the didactic activities and the learning process; it comprises 26 items (4 of which are open questions or requests for explanatory comments). The fourth section investigates personal factors and presents 9 items (4 of which are open questions or requests for explanatory comments). The fifth and last section comprises 20 items (3 of which are open questions or requests for explanatory comments) and investigates the students' opinions about technology-based learning processes and expectations for the future. The complete Italian questionnaire is available at [14].

9.2.2 Participants

The student-set that took part in the survey was composed of 159 elements attending the bachelor's in Education Science at the University of Rome Tor Vergata. Almost all the students were women (98%) attending either first, second and third year of the bachelor course; average age 23.9 years [23.0, 24.8], mode 20; median 30; interval 20–51. The questionnaire realized by means of a Google form has been filled anonymously on February 2021, almost one year after the study described in [4].

9.3 Results

To explore university students' feelings and opinions, as well as the complex causal network that connects the factors investigated in Sections II–V of the survey, we pursued multiple strategies, as in [5–7]. First, we carried out descriptive and univariate analyses (Sect. 9.3.1). For Likert-type response scales, we carried out one-sample t-tests against the midpoint of the scale (5.5 for 10-point scales, 0 for scales ranging from -5 to 5, 0 for the 0–100% scales). The results of the t-tests are reported in Table 9.2. Subsequently, to obtain a bird's-eye view of the factors' relationships, we

Table 9.2 Outcomes of the survey on the students' perception about the factors listed in Table 9.1

Variable	Average	t-test	Average April 2020
University connectivity UC	$M = 6.40$ [6.12, 6.68]	$t(160) = 6.35, p < 0.001$, Cohen's $d = 0.50$	
Technological adequacy of university TAU (TAOE in Ref. 5)	$M = 6.20$ [5.88, 6.52]	$t(155) = 4.32, p < 0.001$, Cohen's $d = 0.35$	$M = 7.22$ [7.09, 7.36]
University reaction capacity URC	$M = 6.51$ [6.25, 6.78]	$t(167) = 7.49, p < 0.001$, Cohen's $d = 0.58$	$M = 7.39$ [7.25, 7.54]
Emergency management EM (UR in Ref. 5)	$M = 6.60$ [6.32, 6.88]	$t(168) = 7.81, p < 0.001$, Cohen's $d = 0.60$	
Operational assistance OA	$M = 5.82$ [5.51, 6.13]	$t(165) = 2.01, p = 0.046$, Cohen's $d = 0.96$	
Teachers' technological preparedness TTP (TTR in Ref. 5)	$M = 6.84$ [6.60, 7.08]	$t(169) = 13.03, p < 0.001$, Cohen's $d = 0.85$	$M = 7.22$ [7.06, 7.38]
Teachers' pedagogical preparedness TPP (TPR in Ref. 5)	$M = 7.08$ [6.83, 7.32]	$t(166) = 12.82, p < 0.001$, Cohen's $d = 0.99$	$M = 7.43$ [7.28, 7.58]
Personal connectivity PC	$M = 7.25$ [6.98, 7.52]	$t(160) = 12.82, p < 0.001$, Cohen's $d = 0.98$	
Individual technological adequacy ITA	$M = 7.62$ [7.38, 7.87]	$t(169) = 16.89, p < 0.001$, Cohen's $d = 1.30$	
Individual technological preparedness ITP	$M = 6.94$ [6.72, 7.20]	$t(168) = 11.01, p < 0.001$, Cohen's $d = 1.00$	
Individual innovation propensity IIP	$M = 7.66$ [7.44, 7.88]	$t(168) = 19.37, p < 0.001$, Cohen's $d = 1.49$	
Self-regulation capability SRC	$M = 7.43$ [7.19, 7.67]	$t(165) = 16.05, p < 0.001$, Cohen's $d = 1.24$	
Political action PA	$M = 5.58$ [5.30, 5.87]	$t(11) = 0.58, p = 0.561$, Cohen's $d = 0.04$	
Societal influence SI	$M = 0.93$ [0.61, 1.27]	$t(171) = 5.64, p < 0.001$, Cohen's $d = 0.44$	

(continued)

Table 9.2 (continued)

Variable	Average	t-test	Average April 2020
Influence of the prevalent opinion in the context IPOC	$M = 0.63$ [0.30, 0.97]	$t(166) = 3.77, p < 0.001$, Cohen's $d = 0.29$	
Influence of familiar environment IFE	$M = 1.20$ [0.75, 1.64]	$t(168) = 5.28, p < 0.001$, Cohen's $d = 0.41$	
Psychological problems PSP	$M = 4.44$ [4.01, 4.87]	$t(162) = -4.90, p < 0.001$, Cohen's $d = -0.38$	
Didactic activity integration DAI	$M = 6.07$ [5.72, 6.42]	$t(140) = 0.58, p = 0.002$, Cohen's $d = 0.27$	
Reproducibility of classroom dynamics (RCD)	$M = 5.35$ [5.00, 5.71]	$t(169) = -0.814, p = 0.417$, Cohen's $d = -0.06$	$M = 5.60$ [5.40, 5.79]
Design activities DA	$M = 5.01$ [4.64, 5.38]	$t(169) = -2.61, p = 0.010$, Cohen's $d = -0.20$	
Collaborative activities CA	$M = 5.26$ [4.91, 5.61]	$t(167) = -0.34, p = 0.181$, Cohen's $d = -0.10$	
Collaborative activities quality CAQ	$M = 6.00$ [5.63, 6.37]	$t(168) = 17.95, p < 0.001$, Cohen's $d = 0.22$	
Collaborative activities percentage CAP	$M = 0.36$ [0.32, 0.40]	$t(168) = 17.95, p < 0.001$, Cohen's $d = 1.38$	
Collaboration with peers CP	$M = 7.22$ [6.97, 7.50]	$t(169) = 12.46, p < 0.001$, Cohen's $d = 0.96$	
Communication with teachers CWT	$M = 5.96$ [5.64, 6.29]	$t(166) = 2.84, p = 0.005$, Cohen's $d = 0.22$	
Communication with peers CWP	$M = 6.80$ [6.49, 7.12]	$t(167) = 8.17, p < 0.001$, Cohen's $d = 0.63$	
Relationships with peers RWP	$M = 0.98$ [0.58, 1.38]	$t(166) = 4.87, p < 0.001$, Cohen's $d = 0.38$	
Relationships with teachers RWT	$M = 0.66$ [0.32, 1.00]	$t(165) = 3.84, p < 0.001$, Cohen's $d = 0.30$	

(continued)

Table 9.2 (continued)

Variable	Average	t-test	Average April 2020
Videoconf usage for lectures VUL	$M = 7.86$ [7.56, 8.16]	$t(170) = 15.33, p < 0.001$, Cohen's $d = 1.17$	
Videoconf usage for discussion VUD	$M = 6.00$ [5.65, 6.35]	$t(170) = 2.86, p = 0.005$, Cohen's $d = 0.22$	
Quizz/test usage QTU	$M = 5.78$ [5.42, 6.14]	$t(170) = 1.52, p = 0.130$, Cohen's $d = 0.12$	
Recorded video usage RVU	$M = 5.44$ [5.04, 5.84]	$t(170) = -0.30, p = 0.762$, Cohen's $d = -0.02$	
Forum usage FU	$M = 5.62$ [5.27, 5.98]	$t(169) = 0.69, p = 0.490$, Cohen's $d = 0.05$	
Dashboard usage DU	$M = 5.69$ [5.33, 6.05]	$t(169) = 1.03, p = 0.306$, Cohen's $d = 0.05$	
Material repository usage MRU	$M = 6.69$ [6.37, 7.02]	$t(168) = 7.24, p < 0.001$, Cohen's $d = 0.56$	
Easyness of use of the integrated didactic EUID (EUOL in Ref. 5)	$M = 1.32$ [1.01, 1.63]	$t(166) = 8.39, p < 0.001$, Cohen's $d = 0.65$	
Usefulness of the integrated didactic UID (UOL in Ref. 5)	$M = 1.90$ [1.58, 2.22]	$t(166) = 11.82, p < 0.001$, Cohen's $d = 0.91$	
Usefulness of technologies UOT	$M = 2.21$ [1.91, 2.51]	$t(169) = 14.52, p < 0.001$, Cohen's $d = 1.11$	
Efficacy of use of the integrated didactic (EID)	$M = 1.69$ [1.37, 2.00]	$t(167) = 10.54, p < 0.001$, Cohen's $d = 0.81$	
Change in efficacy of the evaluation CEE	$M = 1.07$ [0.77, 1.38]	$t(165) = 6.90, p < 0.001$, Cohen's $d = 0.54$	
Motivation toward didactic activities MDA	$M = 0.88$ [0.46, 1.30]	$t(168) = 4.18, p < 0.001$, Cohen's $d = 0.32$	
Personal time management capacity PTMC (TTMC in Ref. 5) (scale $-5, +5$)	$M = 1.33$ [0.94, 1.73]	$t(167) = 6.69, p < 0.001$, Cohen's $d = 0.52$	

(continued)

Table 9.2 (continued)

Variable	Average	t-test	Average April 2020
Improvement in the feeling toward technologies (IFT)	$M = 7.04\ [6.76, 7.31]$	$t(169) = 10.94, p < 0.001$, Cohen's $d = 0.84$	$M = 6.62\ [6.44, 6.79]$
Improvement in technological skills (ITS)	$M = 6.92\ [6.63, 7.22]$	$t(170) = 9.45, p < 0.001$, Cohen's $d = 0.72$	$M = 6.80\ [6.64, 6.96]$
Change in didactic activities (CDA) %	$M = 0.50\ [0.46, 0.54]$	$t(166) = 25.45, p < 0.001$, Cohen's $d = 1.97$	
Change in evaluation modality (CEM) %	$M = 0.39\ [0.34, 0.44]$	$t(162) = 16.66, p < 0.001$, Cohen's $d = 1.30$	
Workload increase (WI) %	$M = 0.30\ [0.26, 0.34]$	$t(165) = 14.88, p < 0.001$, Cohen's $d = 1.15$	
University innovation propensity UIP	$M = 6.87\ [6.62, 7.12]$	$t(169) = 10.73, p < 0.001$, Cohen's $d = 0.82$	
Degree of university e-maturity (UeM)	$M = 6.29\ [6.03, 6.55]$	$t(168) = 5.92, p < 0.001$, Cohen's $d = 0.46$	$M = 7.09\ [6.95, 7.24]$ Before pandemic: $M = 6.83\ [6.72, 6.94]$
Extent to which university should rely on integrated didactic URID	$M = 6.28\ [5.92, 6.64]$	$t(164) = 4.23, p < 0.001$, Cohen's $d = 0.33$	
Sustainability of online education (SOE)	$M = 6.76\ [6.50, 7.02]$	$t(169) = 9.53, p < 0.001$, Cohen's $d = 0.73$	$M = 6.92\ [6.76, 7.07]$
Usefulness of education on integrated didactic UEID (REDP in Ref. 5)	$M = 7.12\ [6.80, 7.49]$	$t(167) = 9.91, p < 0.001$, Cohen's $d = 0.76$	
Intention to use integrated didactic IEID	$M = 6.19\ [5.83, 6.54]$	$t(165) = 3.83, p < 0.001$, Cohen's $d = 0.30$	
Intention to work in smart working IWSW (IWOL in Ref. 5)	$M = 5.75\ [5.37, 6.13]$	$t(167) = 1.29, p = 0.199$, Cohen's $d = 0.09$	$M = 5.61\ [6.37, 5.84]$
Percentage of use integrated didactic PUID %	$M = 0.52\ [0.48, 0.56]$	$t(165) = 25.60, p < 0.001$, Cohen's $d = 1.98$	

employed network analysis to obtain a graph of factors' associations. More specifically, we used the PC algorithm to infer the direction of causality in the graph and to aid interpretation (Sect. 9.3.2) [15].

9.3.1   Descriptive Analysis

Technological context. The *individual technological setting* is quite similar to the one observed one year ago, immediately after the pandemic outbreak [4]. The large majority of the students still use laptops (78%), possibly in parallel with smartphones (36%), tablets (19%) and desktop computers (10%). 41% has a wide or ultra-wide optical fiber connection, 40% has an ADSL connection, about 13% uses the mobile connection and 6% other kinds of connections. It is worthwhile noting, as shown by Table 9.1, that the individual technological setting is considered by most of the respondents more satisfactory than the one provided by the university even though 51% of the students indicate the bandwidth of the connectivity as the main problem encountered during the didactic activities.

The learning ecosystem (i.e. the university): after a first positive impression generated by the readiness with which on March 2020 all universities swapped from face-to-face to online didactic activities the subsequent attempt of managing, although partially, parallel blended learning—i.e. contemporary face-to-face and online teaching—has caused a consistent decrease in the e-Maturity (UeM) [16]. In 2020, in fact, the average value of the UeM of the University jumped from 6.83 to 7.09 while in February 2021 decreased to 6.29. The capability to integrate in presence and at distance activities was judged to be just sufficient, 6.07, but the decrease in the value of UeM seems to have been induced mainly by a limited capability of the university to provide adequate operational assistance (OA $= 5.82$). Another factor that has been considered barely sufficient is the technological adequacy of the University (TAU) which dropped from 7.22 to 6.20. The difference between the indicators related to the technological setting made available by the university and those related to the individual technological setting appears quite significant. Apart from the comparison between the technological adequacy of the university and that of the individuals already mentioned above—6.20 (TAU) versus 7.62 (ITA)—a significant difference has been detected also in the quality of the connectivity: 6.40 (UC) in the case of the university versus 7.25 (PC) in the case of individual connectivity. All this, probably, contributed to the observed difference in the perception about the propensity to innovate: 6.87 in the case of university (UIP) and 7.66 at individual level (IIP). Less significant is the perceived difference about competences, i.e. between the technological preparedness of students and that of teachers: 6.94 (ITP) versus 6.84 (TTP), although TTP is lower than the value, 7.22, measured one month after the pandemic outbreak.

It is important to note that the above negative trends—possibly caused by the attempt to manage in parallel different modalities of participation to the learning activities and, as well, by the uncertainty about the teaching modality to be adopted—do not seem to have influenced the level of collaboration and cohesion between peers, 7.22 (CP). Peer collaboration has been probably very important also to counteract the low level of operational assistence (OA).

External factors. The students express a fairly negative opinion on the action of the politicians, 5.58 (PA). The prevailing opinions within the students' context of reference did not seem to have a significant influence on the respondents' perception, although we detected a slight positive effect: IPOC = 0.63 on a −5/5 scale. The effect of the public opinion seems to be slightly higher: SI = 0.93 on the −5/5 scale.

At personal level, we observed a positive perception about the family environment used by students for the online activities (IFE) (1.20 on the −5/5 scale), while, on average, no significative psychological problems seem to have arisen in most of the respondents: PSP = 4.44/10.

Learning process. As already mentioned above, we detected a barely sufficient evaluation for the capability to integrate face-to-face and online activities (DAI = 6.07). As far as the use of technologies we observed a massive use of videoconference tools to deliver synchronous lessons (VUL = 7.86), while they have been much less used to stimulate discussions (VUD = 6.00). The second most used category of applications is the online repositories to store e-materials (MRU = 6.69), followed at a certain distance by quizzes/tests (QTU = 5.78), message boards (DU = 5.69), forums (FU = 5.62) and recorded videos (RVU = 5.44). The use of collaborative activities has been quite limited (CA = 5.26), even less that of design activities (DA = 5.01). The above outcomes seem to indicate a prevalent use of transmissive approaches but, interestingly, the students did not feel as if the class dynamics were reproduced: RCD = 5.35 (5.60 the value measured previously). The students, in fact, felt that the teaching activities have undergone substantial changes (CDA = 50%) as well as the evaluation processes (CEM = 39%). These perceived changes are accompanied by an impression of greater effectiveness for the didactic activities (CEID = 1.69 on the scale −5/5) and for the evaluation ones (CEE = 1.07). Despite the perception of a 30% increase in the workload (WI), students seem to be able to better organize their time (PTMC = 1.33 on a −5/5 scale). They also felt to have a quite high capacity of self-regulate SRC = 7.43, accompanied by a slight increase in motivation toward the didactic process (MDA = 0.88 on the −5/5 scale).

If from one side we observe an improvement in interpersonal relations—either with teachers, RWT = 0.66 (on the scale −5/5), and among peers, RWP = 0.98 (on the scale −5/5)—from the other the quality of the communication with teachers is somewhat critical: CWT = 5.96. Maybe this value can be partially explained by the limited use of forums and bulletin boards, as well as, by the scarce use of collaborative and design practices that usually involve much dense interactions.

Perceived values. Balancing lights and shadows, after one year, the students' perceptions about distance and parallel blended teaching seem slight positive, as confirmed also by an increase: (a) in the perceived usefulness of technologies (UOT) (2.21 on the −5/5 scale); (b) and of the online/integrated didactic (UID) (1.90 on the −5/5 scale); (c) in the perception of easiness of use (EUID) (1.32 on the $-5/5$ scale). These figures are accompanied by an increase in the individual feeling for the technologies, IFT $= 7.04$ (significantly higher than the one recorded one month after the lockdown, 6.62), and for her/his own technological skills, ITS $= 6.92$ (also higher than the value recorded a year ago, 6.80). The consequence is a more than positive perception about the sustainability of distance and blended learning, 6.76 (SOE), although slightly lower than the value detected immediately after the lockdown, 6.92; a decrease that, as seen above, may be ascribed to the problems encountered by the university in the implementation and management of the parallel blended learning modality.

Future intentions. The overall positive feeling, however, does not seem capable of stimulating in the respondents a desire to work in the next future at distance (in this case as distance educators), IWSW $= 5.75$; probably because the goal of most of them is to get employed in kindergartens. Nevertheless, the need and the relevance of an adequate training in digital pedagogy is widely recognized, 7.12 (UEID).

As regards the future intention of use of online and/or integrated learning for personal training, we observed a more than sufficient value IEID $= 6.19$ and an even slightly higher value for what concern the extent to which the university should rely on the integrated didactic, URID $= 6.28$. As far as its percentage of use, the students suggest as reasonable, on average, 52% (PUID).

9.3.2 Causal Discovery

The search for causal relations returns the picture of Fig. 9.1. It looks like reasonable although it should be interpreted as a tentative one due to the limited number of respondents that tends to cause both the observation of isolated clusters of factors and the impossibility to identify the direction of all the relations.

Let's start by analyzing the clusters of variables, or single variables, not connected to the main structure of the causal network:

- the perceived increase in the workload (WI) seems to be related to the perceived changes in the learning activities (CDA) and evaluation methods (CEM), together with the possible onset of psychological problems (PSP);
- the ability to organize her/his own time more adequately and effectively (PTMC) depends on the increased propensity toward technology (IFT)—which is closely related to the perceived increase in her/his own digital skills (ITS)—as well as on the increased motivation toward online educational processes (MDA);
- political actions (PA), public opinion (SI) and opinions circulating in the students' main context of reference (IPOC) do not seem to have any influence on the students' perception, opinions and future intentions. The family context (IFE) also seems to be irrelevant.

As far as the body of the causal network, from Fig. 9.1 we can see that:

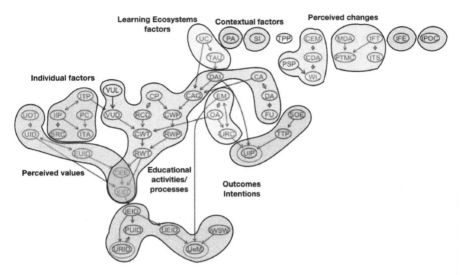

Fig. 9.1 Causal structure of the main variables considered in this study. The red ellipses identify those that can be considered terminal factors of the causal structure

- the individual factors (taken into large consideration by the DOI and UTAUT models) are essentially interrelated with each other; the individual propensity for innovation (IIP) together with the perception of the quality of her/his own Internet connectivity (PC) seem to influence the perception of individual technological adequacy (ITA); in addition, IIP appears to be connected to both the capability of self-regulation (SRC) and the level of individual technological preparedness (ITP); this latter acts as a bridge toward educational activities through the most widely used tools, such as videoconferencing aimed at supporting interaction/collaboration (VUD);
- the factors that characterize the learning ecosystem are divided into two blocks, although both seem to be connected to the level of the ecosystem propensity for innovation (UIP) and to its digital maturity (UeM); the connection for the block that includes the management aspects—i.e. the ability to manage the emergency (EM) and to provide operational assistance (OA)—is more direct and evident; less immediate is the connection for the block that contains the more technological aspects—i.e. the quality of university Internet connectivity (UC) and the adequacy of technologies (TAU)—which to reach UIP must go first through the level of integration of the face-to-face and online learning activities (DAI), then through the block of factors related to the didactic process (light orange area in Fig. 9.1) and, finally, through those related to future expectations/intentions (light gray areas);
- the block of factors related to didactic activities and to the educational process constitutes the backbone of the causal network; three are the factors of particular relevance: (a) the ability to integrate face-to-face and online modalities (DAI); (b)

the use of design-based (DA) and collaborative (CA) activities that well integrate with the use of tools such as forum (FU) and videoconferencing (VUD) used for collaborative purposes; (c) the collaboration and the interactions between peers and with teachers (CP, CWP, CWT, RWP, RWT), as well as the quality of the collaboration (CAQ), factors that are also linked to the ability to reproduce the classroom dynamics (RCD).

Downstream of this block, as terminals of the causal network, we find the effectiveness of the integrated/blended learning activities (EID) and that of the evaluation process (EEC). It is really interesting to note how the factors related to the perceived values—considered mainly, although not exclusively, by the TAM model—converge on EID and play a supporting but not essential role to the development of the causal network, as already observed in [7].

The perception of effectiveness is the bridge toward the future intentions/expectations represented by the individual intention to use integrated learning processes (IEID)—together with their percentage of use (PUID)—and the expectations on how much the university should rely on them (URID). The relevance of an adequate training on technologies for integrated learning activities (UEID) together with the availability to work in future at distance (smart working—IWSW) and the ability to manage the emergency (EM) influence the e-maturity level of the learning ecosystem (UeM).

9.4 Discussion and Conclusions

Combining the evidence of Sects. 9.2 and 9.3 we can state that the organization and delivery of parallel blended learning (contemporaneity or alternation of face-to-face and online didactic activities), adopted as main strategy to deliver educational processes during the second phase of the pandemic, would have required a more robust infrastructure with respect to the case of learning activities carried on exclusively online. A more timely and adequate consideration of such criticality might have determined a more positive students' perception of the ecosystem, of its propensity to innovate and of its ability to manage the emergency, as clearly evidenced, by contrast, by the higher value assigned to the individual technological setting.

We can also speculate that the overall positive perception of students about online and integrated learning could have been even more positive if the educational processes would have included more design-based and collaborative activities. Collaboration and interaction between peers and with teachers, in fact, is the backbone of the casual network shown in Fig. 9.1.

Students seem to consider the experience of online or integrated learning in a very pragmatic way and base their opinions on the quality of the didactic process and its effectiveness that, obviously, is also determined by the adequacy of the technological setting that should guarantee an optimal integration of the various didactic strategies and modalities. Other factors, e.g. the contextual ones, do not seem to significantly influence their opinion, nor their future intentions/availability to use online modality for their training. All this seems to indicate that the physical presence when not strictly needed (laboratory activities, internships, etc.) is more a goal of the University than of the students (apart from their wish to experience a denser social interaction).

Individual factors seem to be confined within the personal sphere and to be related to the ability to optimize her/his own time and to support a feeling of greater individual adequacy in participating in online processes.

Overall, the structural organization that we can derive from the causal network of Fig. 9.1 is that shown in Fig. 9.2. Even though in this work we are considering the opinions of students, the structure of Fig. 9.2 tends to confirm the main causal chain observed in the MAETI model [7] that was obtained on the basis of the opinions expressed by university teachers. In fact both originate from the organizational-technological setting and converge on the future intentions passing through aspects related to the didactic process. It is also confirmed the side relevance of the contributions provided by individual factors and by the perceived values, apart from the effectiveness of the didactic activities. In conclusion, we can state that the backbone of the MAETI is confirmed although further investigations are needed to better identify the role of and confirm the relations among the various blocks of factors (e.g. for the case of the perceived sustainability).

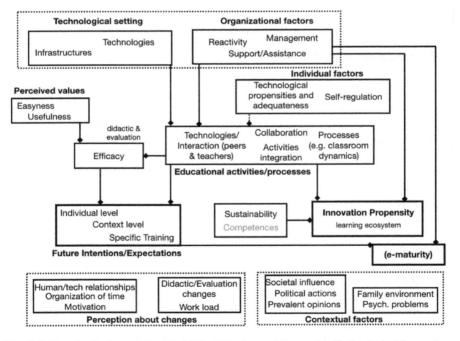

Fig. 9.2 The sketch of a tentative Model for Attitude to get Engaged in Technological Innovation (MAETI) for the present case study

The outcomes of the present work seem to suggest the need to rethink critically the models of technology innovation and acceptance, at least for complex and systemic changes like the one that we have all experienced during this last year of pandemic. At the same time, the above considerations also highlight the limits that would affect any attempt to compare the outcomes of technology acceptance surveys conducted before and during the pandemic.

References

1. UNESCO: https://en.unesco.org/themes/education-emergencies/coronavirus-school-closures (2020)
2. Sahin, I., Shelly, M. (eds.): Educational Practices During the COVID-19 Viral Outbreak: International Perspectives. ITES Organization (2020)
3. ASLERD Ed: Learning and learning ecosystems in the time of COVID-19. Interact. Des. Archit. J. (IxD&A) 46 (2020)
4. Giovannella, C.: Effect induced by the Covid-19 pandemic on students' perception about technologies and distance learning. In: Pedagogical Approaches, Ludic and Co-Design Strategies & Tools Supporting Smart Learning Ecosystems and Smart Education. Springer Verlag (2020)
5. Giovannella, C., Passarelli, M.: The effects of the Covid-19 pandemic seen through the lens of the Italian university teachers and the comparison with school teachers' perspective. Interact. Des. Archit. J. (IxD&A) **46**, 120–136 (2020)
6. Giovannella, C., Passarelli, M., Persico, D.: Measuring the effects of the Covid-19 pandemic on Italian learning ecosystems at the steady state: the school teachers' perspective. Interact. Des. Archit. J. (IxD&A) **45**, 264–286 (2020)
7. Giovannella, C., Passarelli, M., Alkhafaji, A.S.A., Peña Pérez Negrón A.: A comparative study on the effects of the COVID-19 pandemic on three different national university learning ecosystems as bases to derive a Model for the Attitude to get Engaged in Technological Innovation (MAETI). Interact. Des. Archit. J. (IxD&A) **47** (in print)
8. Rogers, E.M.: Diffusion of Innovations, 5th edn. Free Press, New York (2003)
9. Leoste, J., Heidmets, M., Ley, T., Stepanova, J.: Classroom innovation becoming sustainable: a study of technological innovation adoption by Estonian primary school teachers. Interact. Des. Archit. J. (IxD&A) **47** (in print)
10. Davis, F.D.: A Technology Acceptance Model for Empirically Testing New End-User Information Systems: Theory and Results. Massachusetts Institute of Technology (1985)
11. DePietro, R., Wiarda, E., Fleischer, M.: The context for change: organization, technology and environment. In: Tornatzky, L.G., Fleischer, M. (eds.) The Process of Technological Innovation, pp. 151–175. Lexington Books, Lexington, MA (1990)
12. Venkatesh, V., Morris, M.G., Davis, G.B., Davis, F.D.: User acceptance of information technology: toward a unified view. MIS Q. **27**(3), 425–478 (2003)
13. Ley, T., Maier, R., Thalmann, S., et al.: A knowledge appropriation model to connect scaffolded learning and knowledge maturation in workplace learning settings. Vocat. Learn. **13**, 91–112 (2020)
14. https://docs.google.com/forms/d/e/1FAIpQLSeFROFm7xI3sLuKVsaReMWyc3cIj55u1x7q P2ReQmrS5Z3aEQ/viewform
15. Kalisch, M., Maechler, M., Colombo, D., Maathuis, M.H., Buehlmann, P.: Causal inference using graphical models with the R package pcalg. J. Stat. Softw. **47**(11). 1–26 (2012). http://www.jstatsoft.org/v47/i11/
16. Sergis, S., Sampson, D.G.: From teachers' to schools' ICT competence profiles. In: Theorizing Why in Digital Learning: Opening Frontiers for Inquiry and Innovation with Technology, pp. 307–327. Springer (2014)

Chapter 10
A Snapshot of University Students' Perceptions About Online Learning During the Covid-19 Pandemic

Stavros A. Nikou ⓘ

Abstract Gaining a better understanding of how the Covid-19 pandemic has affected students' attitudes towards online education is important in order to pave the way from education disruption to education recovery. The current study aims to investigate how Covid-19 pandemic has changed University students' perceptions about online education. The study used a localised version of a questionnaire developed by the Association for Smart Learning Ecosystem and Regional Development on different aspects of distance education. Participants were 90 University students. The results emerging from the investigation demonstrate a positive overall attitude of University students about online distance education, with relatively high levels of perceived ease of use, perceived usefulness, perceived interest and digital competences increase and a preference towards blended modes of delivery. The findings can help education professionals to better plan and design future online courses in the post-Covid-19 landscape.

Keywords Distance education · Online learning · Perception about technologies · Smart learning ecosystems · Covid-19 pandemic

10.1 Introduction

During 2020, educational institutions around the world have been experiencing a prolonged lockdown due to the Covid-19 pandemic with the educational delivery to be shifted from face-to-face to online. According to UNESCO, 1.3 billion students were affected by school closures in 195 countries—from pre-primary to higher education [1]. However, digital technologies managed to ensure the continuity of learning worldwide [2]. This shift to online teaching, described as Emergency-Remote Teaching (ERT) [3], has raised a debate on whether or not online learning should continue to play a predominant role in the post-Covid-19 era. During 2021, a year after the pandemic hit, close to half the world's students are still affected by

S. A. Nikou (✉)
School of Education, University of Strathclyde, Glasgow, UK
e-mail: stavros.nikou@strath.ac.uk

© The Author(s), under exclusive license to Springer Nature Singapore Pte Ltd. 2022   123
Ó. Mealha et al. (eds.), *Ludic, Co-design and Tools Supporting Smart Learning Ecosystems and Smart Education*, Smart Innovation, Systems and Technologies 249,
https://doi.org/10.1007/978-981-16-3930-2_10

partial or full school closures [4] and educational institutions around the world, due to social distance restrictions that continue to be in place, continue to deliver online classes eliminating face-to-face instruction. As educational systems work to build resilience and adaptability towards technology [5], online and blended learning are in the centre of discussions worldwide. There are many issues to consider during this journey from disruption to recovery. Ensuring access to information technologies for all children, mitigating the impact of learning losses during the pandemic, better understanding the interplay between technology and pedagogy, developing appropriate digital pedagogical resources, are a few of these issues [5].

There have been numerous efforts to investigate the impact of Covid-19 pandemic on the educational community worldwide [6, 7]. One of these efforts has been initiated by the Association for Smart Learning Ecosystem and Regional Development (ASLERD). ASLERD has developed questionnaires to investigate how Covid-19 pandemic has affected University and High schools teachers' and students' perceptions about distance education. The surveys gather current feelings and opinions on some aspects of the distance learning experience from the University teachers and students perspectives [8–10].

In the same context, the current study aims to investigate aspects of distance learning during the Covid-19 pandemic from the perspective of University students. The study has been based on a modified and localised version of the ASLERD questionnaire for students and has been carried out in a UK University. This work in progress is aiming to provide a snapshot of University students' perceptions about online learning during the Covid-19 pandemic. The following sections describe the study methodology, the data analysis and results and also conclusions and discussions.

10.2 Methods

10.2.1 Participants—Procedures

The study has been conducted during November 2020, after having received the required ethics approval. The researchers contacted potential participants and informed them about the aims of the research. Participation in the study was voluntary. Students who agreed to participate were asked to give their informed consent. The email to the prospective participants included the participant information sheet, the consent form and a link to the online Qualtrics questionnaire.

The respondents in the questionnaire were 90 University students from a Department of Education in a UK University. There were 80 females (89%) and 10 males (11%). 63 students (70%) were undergraduate while 27 students (30%) were postgraduate. There were 46 students below 22 years old (51%), 20 students between 23 and 30 years old (22%) and 24 students above 30 years old (27%).

Regarding participating students' previous experience in online learning, 73% of the students self-reported a rather low previous experience (levels 1–5 in a 10-point Likert-type scale) and only 27% reported higher levels of previous online learning experience (levels 6–10 in the same scale). Overall, the previous experience with online learning of the participants was quite low, $M = 4.20$ (0.29) on a 10 point Likert-like scale (1–10).

Regarding student engagement with online learning, 23% of the students reported that they were engaged more than four hours a day, 43% of the students were engaged between two hours and four hours a day, 13% between an hour and two hours, 14% between 30 min and an hour and 7% less than 30 min.

10.2.2   Questionnaire

The study used a modified and localised version of the original questionnaire developed by the Association for Smart Learning Ecosystems and Regional Development (ASLERD) about student attitudes on distance learning. ASLERD has initiated a global research to investigate the effect of pandemic on education. The ASLERD questionnaire is focused on students' feeling about online activities, the change in their perceptions about online learning and their future expectations related to online education [8]. The title of the original questionnaire is "Me and the distance learning—university student questionnaire".

Our slightly modified version of the aforementioned ASLERD questionnaire is a subset of the original. The questionnaire is composed of five introductory questions (gender, age, educational level, previous experience with online learning, time engaged in online learning per day) and twenty questions related to students' opinions on distance learning as this has been experienced during the Covid-19 pandemic. The tool has been validated through pilot testing. Questions were both closed type (multiple-choice type, Likert-type) and open type as well where students could provide their comments to explain their answers to the close-type questions (multiple choice and numerical linear scale). The filling time of the questionnaire was less than 10 min if students answered only multiple-choice questions, checkboxes and linear scale questions and about 20 min if they answered also open questions. The questionnaire was answered anonymously.

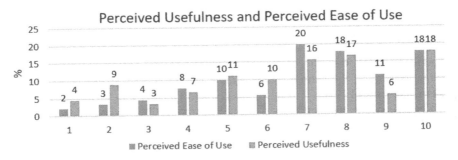

Fig. 10.1 Perceived usefulness and perceived ease of use

10.3 Results on Students' Perceptions About Online Learning

10.3.1 Perceived Usefulness and Perceived Ease of Use

Most students reported that they perceive online learning as being easy and useful. On a 10-point Likert scale the average perceived ease of use and perceived usefulness were reported as being quite high, e.g. 6.44 ± 0.28 for the perceived ease of use and 6.97 ± 0.25 for the perceived usefulness. 56% of the students perceived online learning as easy and 47% of the students perceived online learning as useful (both variables ranked above 7 on a 10-point scale), as Fig. 10.1 shows. Table 10.1 shows t-tests performed against 5.5 the midpoint of the scale. Results have shown that the mean is significantly greater than the midpoint with $t(89) = 5.785$, $p < 0.001$ and $t(89) = 3.325$, $p < 0.001$ and the corresponding effect sizes to be rather high for the perceived ease of use (0.61) and medium for the perceived usefulness (0.35) respectively.

10.3.2 Studying Online

The average perceived increase in educational working load online (as Fig. 10.2a shows) was reported as being close to the midpoint of the 10-point Likert scale with a value of 4.58 ± 0.33. Most of the students (59%) believe that online learning cannot heavily increase (values 1–5 in the 10 point scale) their working load, $t(86) = -2.734$, $p < 0.01$.

Regarding the impact of online learning on self-organising learning activities and time management (Fig. 10.2b), 21% of the students do not consider that online learning can have an impact, 24% believe that it can have a negative impact, while 55% believe that it can have a positive impact and improve their time management and organisation. The average value is 5.7 ± 0.30.

Table 10.1 Survey results

Variable	Mean (std. error)	t-test	Effect size (Cohen d)
Previous online experience	4.20 (0.29)	$t(89) = 4.409***$	0.49
Perceived usefulness	6.44 (0.28)	$t(89) = 3.325***$	0.35
Perceived ease of use	6.97 (0.25)	$t(89) = 5.785***$	0.61
Increase working load	4.58 (0.33)	$t(86) = -2.734**$	0.29
Self-organisation	5.70 (0.30)	$t(86) = 0.670$	0.07
Interest in learning technologies	4.32 (0.32)	$t(80) = -3.625***$	0.41
Online competencies	4.92 (0.27)	$t(80) = -2.051*$	0.22
Change in educational experience	4.80 (0.30)	$t(79) = -2.309*$	0.26
Missed f2f classes	6.81 (0.38)	$t(78) = 3.467***$	0.39
University to continue e-learning	5.62 (0.27)	$t(79) = 0.457$	0.05
Working future as remote	4.85 (0.36)	$t(75) = -1.788$	0.20

$*p < 0.05, **p < 0.01, ***p < 0.001$

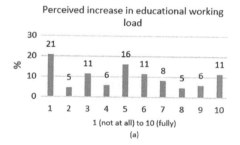 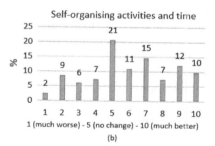

Fig. 10.2 Perceptions on educational working load (**a**) and self-organising activities and time (**b**)

10.3.3 Interest and Competency in Educational Technologies

The average value of the perceived increase in interest in learning technologies is 4.32 ± 0.32. 27% of the students self-reported a full increase in learning technologies, while 33% reported almost no impact (values 4–6 in the 10-point scale). The population mean is significantly different from the midpoint of the 10-point scale $t(80) = -3.625, p < 0.001$ indicating a positive impact on the interest in learning technologies (Fig. 10.3a).

Regarding the impact on educational technology competencies, 64% of the students self-reported a strong positive impact (values 6–10) and only 36% reported

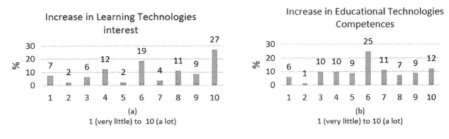

Fig. 10.3 Interest in learning technologies (**a**) and educational competencies (**b**)

a rather lower positive impact (values 1–5). The mean value is 4.92 ± 0.27 with $t(80)$ $= -2.051, p < 0.05$ (Fig. 10.3b).

10.3.4 Perceptions About Educational Technologies

Two important aspects that most students reported as a positive effect of online learning (Fig. 10.4a) are (i) the development of students' own digital identity (58%) and (ii) learning to work autonomously and develop their self-regulation (54%). Many students (39%) reported that learning efficiency is another improvement that online education can offer while learning effectiveness and designing and managing the learning process are important potential improvements as well (29%). Also online learning improves the quality of learning experience (28%) and facilitates interactions development.

Educational technologies are considered useful (Fig. 10.4b) by students because they facilitate content sharing (76%), delivery of transmissive non-interactive lessons (e.g. video clips) (69%) and assignment of asynchronous tasks (61%). Communication with teachers (50%), delivering interactive lessons (40%) and carrying out exercises online (40%) are three aspects of the synchronous mode that educational technologies can support. 35% of student responses credit collaboration and team working.

61% of the participating students reported that their idea of educational experience did not change much (values 1–5) while 39% self-reported a considerable change. The population mean is 4.80 ± 0.30, differing from the midpoint of the scale, $t(79)$ $= -2.309, p < 0.05$. Only 7% reported that their idea of educational experience has fully changed (Fig. 10.4c).

Many students (65%) missed face-to-face classes (reported a score more than 5 in 10-point Likert scale). In this scale, the average perception of having missed face-to-face classes is 6.81 ± 0.38 and the sample mean is significantly greater than the midpoint of the scale, $t(78) = 3.467, p < 0.001$ as Fig. 10.4d shows.

10.3.5   Future Expectations

Most participants (48%) would prefer to continue having a blended learning approach, while 30% would be in favour of face-to-face and 20% of online teaching, as Fig. 10.5a shows.

Students (77%) prefer lectures to be offered online comparing to face-to-face (23%). For sessions that require more interactivity, students prefer to have them face-to-face instead of online, e.g. for the seminars/tutorials 73% prefer online versus 27%

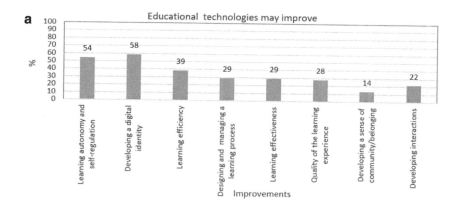

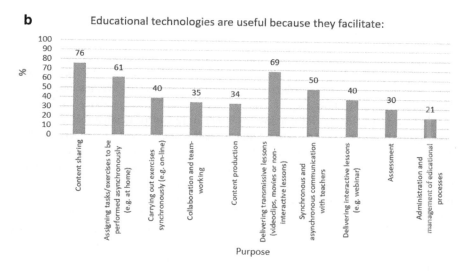

Fig. 10.4   a Perceptions on the improvements of educational technologies. **b** Perceptions on the usefulness of educational technologies. **c** Change of students' idea of educational experience. **d** How much students missed face-to-face classes

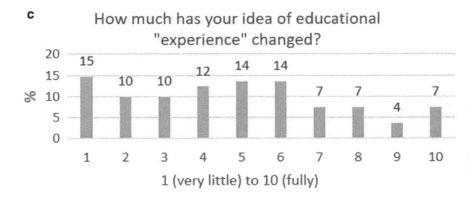

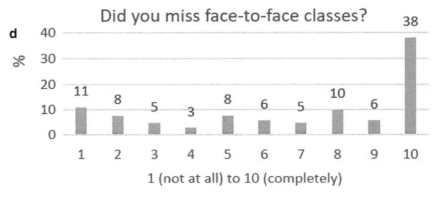

Fig. 10.4 (continued)

face-to-face and for the practical sessions 81% prefer online versus 19% face-to-face. For the assessments, the preferred delivery mode is online (60%) versus face-to-face (40%), as Fig. 10.5b shows.

Many students (21%) did not answer either in favour or against the idea of the University to continue using online educational activities, as Fig. 10.5c(a) shows. 46% of the students are in favour of the University to continue using online educational activities while 33% are rather unsure. On a 10-point Likert scale the average preference is 5.62 ($\pm$ 0.27) while performing t-tests against the midpoint of the scale, the preference towards online educational activities has been reported with a positive but rather low effect size (0.25). Also, students did not seem to have a clear preference on whether they would prefer their future working career to be online, as Fig. 10.5c(b) shows. 49% reported a score lower than the midpoint while 51% reported a positive preference. The mean does not significantly differ from the midpoint of the 10-item Likert scale, $t(75) = -0.401, p > 0.05$.

a

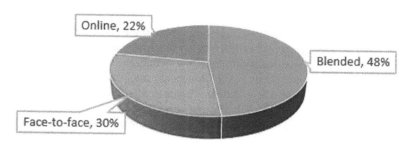

b

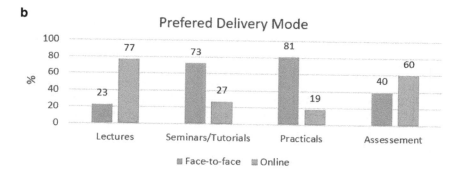

c

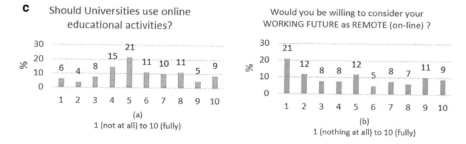

Fig. 10.5 a Students' general preference for the teaching delivery mode. **b** Students' preference for the delivery mode of lectures, seminars, practicals, assessment. **c** Students' preference on the continuance of online learning (**a**) and online working (**b**)

10.4 Conclusions and Discussions

The current study has been conducted during a period of a global educational disruption due to the Covid-19 pandemic. The purpose of the study is to investigate aspects of University students' perceptions on distance learning. The study is based on a modified and localised version of the ASLERD questionnaire "Me and the distance learning—university student questionnaire".

While educational institutions around the world plan to move from education disruption (due to the Covid-19 pandemic) to education recovery, there are different issues that need to be considered. To ensure access to information technologies for all children, to mitigate the impact of learning losses during the pandemic, to better understand the interplay between technology and pedagogy, and to develop appropriate digital pedagogical resources and appropriately integrate online technologies to instructional design [5], are a few of them. In this process, students' attitudes and perceptions about educational technologies are very important factors to be considered. The current small-scale study is aiming to contribute towards this direction. It is part of our larger project that investigates the impact of this online transition on teachers' and students' attitudes on technology-enhanced learning [11].

The results emerging from the current investigation (albeit preliminary) demonstrate a positive overall attitude of University students about online distance education. Students perceived online learning as easy to use and useful. Perceived usefulness and perceived ease of use are the most important determinates of technology acceptance [12]. These two main technology acceptance variables have been used in a plethora of studies investigating students' perceptions of information technology [13]. The same variables have been used in other similar studies that investigated the impact of the SARS-CoV2 pandemic in education [9]. Students indicated that online learning had a positive impact on their interest in learning technologies. Despite the challenges associated with online learning due to the complexity of technological variations [14] students self-reported that their competencies in educational technologies have been increased. Regarding the benefits that technologies can offer to improve learning activities, students acknowledged the positive impact of educational technologies in working effectively and efficiently. There is a variety of strategies used in online teaching, and educational technologies are useful in both synchronous and asynchronous modes of educational delivery, which is in line with other recent studies [15]. However, the correlation between adoption of digital technologies and student efficiency may need further investigation [16]. Students answered in a conservative manner on whether or not their educational experience has been considerably changed and improved, since the majority of them have missed their face-to-face classes. This is confirmed by the fact that they have indicated a clear preference for a future blended learning approach. Students prefer lectures to be offered online while for sessions that require more interactivity they would prefer to be face-to-face. Students seem to be ready for adopting online or blended learning processes [8].

The findings of this small-scale work-in-progress study can contribute to previous investigations on student attitudes about educational technologies and how Covid-19 pandemic has affected their perceptions about distance education [8, 9]. However, the results should be interpreted with caution as they are preliminary and the study has limitations. The small sample size is one of the study limitations. Another limitation may be the study demographic from the particular study programme. Future work is aiming to involve a larger number of participants from different study backgrounds.

Acknowledgements Authors would like to thank the students who participated in the research.

References

1. UNESCO: 1.3 billion learners are still affected by school or university closures, as educational institutions start reopening around the world, says UNESCO. https://en.unesco.org/news/13-billion-learners-are-still-affected-school-university-closures-educational-institutions (2020). Accessed 8 Apr 2021
2. UNESCO: Education: from disruption to recovery. https://en.unesco.org/covid19/educationresponse. Accessed 8 Apr 2021
3. Hodges, C.B., Moore, S., Lockee, B.B., Trust, T., Bond, M.A.: The difference between emergency remote teaching and online learning. EDUCAUSE Rev. (2020)
4. UNESCO: 1.37 billion students now home as COVID-19 school closures expand, ministers scale up multimedia approaches to ensure learning continuity says UNESCO. https://en.unesco.org/news/137-billion-students-now-home-covid-19-school-closures-expand-ministers-scale-multimedia (2020). Accessed 8 Apr 2021
5. OECD: Coronavirus Special Edition: Back to School, Trends Shaping Education Spotlights, No. 21. OECD Publishing, Paris (2020). https://doi.org/10.1787/339780fd-en
6. Di Pietro, G., Biagi, F., Costa, P., Karpiński Z., Mazza, J.: The likely impact of COVID-19 on education: reflections based on the existing literature and international datasets, EUR 30275 EN. Publications Office of the European Union, Luxembourg (2020). https://doi.org/10.2760/126686, JRC121071
7. IAOU: Regional and National Perspectives on the Impact of COVID-19 on Higher Education. International Association of Universities. https://www.iau-aiu.net/IMG/pdf/iau_covid-19_regional_perspectives_on_the_impact_of_covid-19_on_he_july_2020_.pdf (2020). Accessed 8 Apr 2021
8. Giovannella, C.: Effect induced by the Covid-19 pandemic on students' perception about technologies and distance learning. In: Mealha, Ó., Rehm, M., Rebedea, T. (eds.) Ludic, Co-Design and Tools Supporting Smart Learning Ecosystems and Smart Education. Smart Innovation, Systems and Technologies, vol. 197. Springer, Singapore (2021). https://doi.org/10.1007/978-981-15-7383-5_9
9. Giovannella, C., Passarelli, M., Alkhafaji, A., Negrón, A.P.P.: A Model for the Attitude to get Engaged in Technological Innovation (MAETI) derived from a comparative study on the effects of the SARS-CoV2 pandemic seen through the lens of the university teachers of three different national learning ecosystems: Iraq, Italy and Mexico (2020)
10. Giovannella, C., Passarelli, M., Persico, D.: Measuring the effects of the Covid-19 pandemic on Italian Learning Ecosystems at the steady state: the school teachers' perspective. Interact. Des. Archit. J. (IxD&A) **45**, 264–286 (2020)
11. Nikou, S.A.: Web-based videoconferencing for teaching online: continuance intention to use in the post-COVID-19 period. Interact. Des. Archit. J. (IxD&A) **47**, 123–143 (2021)
12. Davis, F.D.: Perceived usefulness, perceived ease of use and user acceptance of information technology. MIS Q. **13**(3), 319–340 (1989)
13. Marangunić, N., Granić, A.: Technology acceptance model: a literature review from 1986 to 2013. Univ. Access Inf. Soc. **14**, 81–95 (2015)
14. Rasheed, R.A., Kamsin, A., Nor Aniza Abdullah, N.A.: Challenges in the online component of blended learning: a systematic review. Comput. Educ. **144**, 103701 (2020)
15. Simpson, J.C.: Distance learning during the early stages of the COVID-19 pandemic: examining K-12 students' and parents' experiences and perspectives. Interact. Des. Archit. J. IxD&A **46**, 29–46 (2020)

16. Lacka, E., Wong, T.C., Haddoud, M.Y.: Can digital technologies improve students' efficiency? Exploring the role of virtual learning environment and social media use in higher education. Comput. Educ. **163**, 104099 (2021)

Chapter 11
College Students' Blended Online Examination Acceptance During the COVID-19 Epidemic

Xin-yu Jiang, Tiong-Thye Goh, and Meng-jun Liu

Abstract To meet the assessments requirements during the COVID-19 epidemic, many schools adopted the large-scale blended online examination, in which teachers invigilate through online video and students answer questions with pen and paper. Exploring the factors that influence students' acceptance of the blended online examination will help the exam service understand the characteristics of students' adoption and provide better support for staffs. It will help teachers and parents to assist students to take the blended examination and maintain the traditional exam atmosphere that will facilitate students' learning performance and effectiveness. In this study, a questionnaire survey and structural equation method were adopted to explore the influence of perceived ease of use, perceived usefulness, social presence, place presence, and social influence on learners' acceptance of blended online examination. Data analysis of 760 college students who underwent a blended online examination showed that perceived ease of use had a significant negative direct impact on exam acceptability and overall had a significant negative impact. Perceived usefulness, social presence, and social influence have significant positive effects on exam acceptability and social presence and social influence also have significant positive effects indirectly by influencing perceived usefulness. Finally, the limitations of this study are discussed, and the implications and future direction are put forward.

Keywords Blended online examination acceptance · Technology acceptance model · Social presence · Social influence

11.1 Introduction

In the early COVID-19 pandemic, when students could not start school as planned, the Ministry of Education initiated the "Suspending Classes Without Stopping

X. Jiang · M. Liu (✉)
School of Education, Hubei University, Wuhan, China
e-mail: lmj_whu@163.com

T.-T. Goh
School of Information Management, Victoria University of Wellington, Wellington, New Zealand

© The Author(s), under exclusive license to Springer Nature Singapore Pte Ltd. 2022   135
Ó. Mealha et al. (eds.), *Ludic, Co-design and Tools Supporting Smart Learning Ecosystems and Smart Education*, Smart Innovation, Systems and Technologies 249,
https://doi.org/10.1007/978-981-16-3930-2_11

Learning." Education authorities provided online learning resources and support services for teachers and students to ensure large-scale online teaching. Online teaching not only ensures the safety and health of nearly 300 million teachers and students in China but also relieves social anxiety and panic and maintains social stability. The examination is an assessment of student's mastery of course content, teachers' teaching effect, teaching quality, and other aspects. To exam learners' learning outcomes, many schools have adopted the blended online examination models, which is a combination of the traditional classroom exam model embedded in an online environment—invigilated by teachers via video conferencing, and students take the pen and paper exam remotely and submitting online. As the first large-scale online exam, this is not only an exam for students but also the education system, school team, and family. Especially, in the blended online examination, there is no teacher supervision, and the scene atmosphere of unified examination requires not only the self-control ability required by learners themselves but also their adaptability and acceptance of the blended online examination. To explore the factors that influence learners' acceptance of online examination is helpful for online exam platforms to understand the characteristics of students' adoption and provide better services. It is helpful for teachers and parents to guide students to take the online examination and maintain the exam atmosphere to ensure learners' learning performance and learning effectiveness.

The online examination has the characteristics of large scale, low cost, high efficiency, and flexibility. When studying learners' different experiences of online exams and traditional classroom exams, Myyry and Joutsenvirta [1] investigated the aspects of exam preparation, question-answering, and learning experience, especially focusing on what learning strategies students adopt in online exams and whether they regard online exams as real-exam situations. Stowell and Bennett [2] found that the online exam environment can reduce test anxiety of students who are highly anxious in the traditional exam. Existing research frequently focuses on the differences between the online exam and traditional exam and the effectiveness of the online examination system design and implementation. This study will investigate the influencing factors affecting learners' willingness to accept the online exam. The study will construct a theoretical research model to investigate the impact of learners' blended online examination acceptance focusing on online learners from a university. A survey was conducted from July 28th to 31st, 2020 to collect data for statistical analysis and to explore the factors affecting learners' blended online examination acceptance. The results will form the online examination policy and provide a theoretical reference for operationalizing the university-wide online exam system and further enhance our knowledge on the influencing factors of online exam acceptance during the COVID-19 pandemic.

11.2   Research Model and Hypotheses

11.2.1   The Technology Acceptance Model (TAM)

Technology acceptance model (TAM) is a research model proposed by Davis et al. [3] based on the Theory of Reasoned Action, combining the Self-Efficacy Theory and the Expectancy Confirmation Theory, which can effectively predict people's intention and behavior of using information technology [4, 5]. Studies have found that in online teaching and learning environments, learners' perception of technology is a key factor for success in online learning, and useful and easy-to-use learning tools can promote learners' engagement and satisfaction [6, 7]. Akour [8] constructed a mobile learning acceptance model based on the TAM to explain the important factors of first-year students' acceptance of mobile learning, including student readiness, accessibility, service quality, and school support. Joo et al. [9] used the TAM to investigate the structure of the relationship between the perceived level of the existence of online learning tools, perceived usefulness, perceived ease of use, learners' satisfaction, and persistent, and the results showed that perceived ease of use and perceived usefulness have a significant role for the learners' satisfaction. In this study, perceived ease of use represents learners' perceived ease of using information technology to carry out the online exam, perceived usefulness indicates the learner's identification with information technology to improve the online exam experience and effectiveness, and behavioral intention is expressed as acceptance, that is, learners' acceptance to adopt and use the blended online examination mode. By perceived ease of use and perceived usefulness influence on learners blended online examination acceptance, learner before the examination, must be familiar with the system function, process and operation in advance, the lack of function, process trivial and operation failure can lead to learners produce resistance, even in the process of online examination is likely to influence the learner's exam, hinder the online exam process smoothly. Therefore, the following hypotheses are proposed:

H1a: Perceived ease of use positively affects learners' blended online examination acceptance.

H1b: Perceived usefulness positively affects learners' blended online examination acceptance.

H1c: Perceived ease of use positively affects perceived usefulness.

11.2.2   Social Presence

Social presence was defined as "the significance of others in interaction and the resulting significance of interpersonal relationship" [10]. Garrison [11] later put forward the theory of Community of Inquiry. Garrison believed that meaningful learning takes place through the interaction of the three core elements of the online learning community, namely social, teaching, and cognitive presence. Social presence refers to the subject's perception of the existence of others, and the individual is in the process of communication as a "real man" [12]. As a kind of virtual experience, social presence was shown to have a positive impact on the behavioral intention of network users. Gunawardena [13] and Palloff and Pratt [14] found through the study of online learning environments that social presence contributed to the formation of a positive learning experience and the improvement of learning satisfaction. And social presence affects online learning emotion and is a significant positive predictor of perceived learning, learning engagement, learning satisfaction, and learning intention [15, 16]. Some studies also showed that social presence has a certain impact on users' perceived usefulness [17]. In addition to the physical distance from the school classroom to the examination room to the family life or the environment, there is also the psychological distance between teachers and students from a strong zero-distance relationship into a weak long-distance relationship. The perception and experience of examination at such a distance are often lonely, thus reducing the learners' insistence on examination. Social presence is not a simple interaction, but a continuous process of social interaction and cognitive communication, from the absence to low level of psychological participation to a high level of behavioral expression [18]. By increasing the sense of presence in the online exam, it can help reduce the loneliness in the remote examination environment and help the students in the state of separation form a sense of belonging and identity. The following hypotheses are proposed:

H2a: Social presence positively affects learners' blended online examination acceptance.

H2b: Social presence positively affects perceived usefulness.

11.2.3 Place Presence

In a virtual environment, the sense of presence (also known as a physical sense of presence) is mainly used to refer to an individual's "sense of presence" in a specific virtual environment. Sheridan [19] defined "presence" as a subjective and psychological feeling. His definition includes three aspects: the feeling of "presence," the individual's response to "presence" as real or present things, and the individual's memory of the environment as "site," just like in real life. Bulu [20] found that learners who feel real in a virtual environment tend to regard the environment as more personal and socialized, and the place presence in the virtual environment has a positive impact on system satisfaction. Due to the outbreak of COVID-19, different from the traditional exam mode, students take exams in the same realistic environment, and students complete the learning tasks and exams alone, and blended online examination is a combination of the situational virtual world environment and real tasks. Through network video conferences, it provides students with a subjective and psychological feeling of invigilation by teachers and students in the same situation in the virtual network environment. The subjective and psychological feeling of students in the same situation of the exam is simulated, which is expected to improve the acceptance of students in the blended online exam. The following hypotheses are proposed:

H3: Place presence positively affects learners' blended online examination acceptance.

11.2.4 Social Influence

Social influence refers to the degree to which a specific person or organization influences learners to take the online examination, which mainly comes from the people around, mass media, and the government's regulations on the industry. Hossain et al. [21] explored the factors influencing employees' behavioral intention of e-learning in the public sector based on the UTAUT model, and the results showed that social influence would significantly affect users' behavioral intention of e-learning. Wu and Zhang [22] confirmed that social influence in online learning plays an important predictive role in college students' willingness to continue. Besides, some studies also proved that social influence can affect the perceived usefulness of learners to some extent. On the willingness to continue using MOOCs, Wu and Chen [23] proved that social recognition and social influence play an important role in predicting the willingness to continue, and perceived usefulness is a significant intermediary between social recognition and social influence on the willingness to continue. The social influence of this study mainly includes the ministry of Education's advocacy of "suspension of classes without suspension," the promotion of students' timely completion of teaching tasks through online learning during the epidemic period, and the proposal and organization of schools and teachers for students to take the blended online examination to complete teaching inspection and assessment (Fig. 11.1). The following hypotheses are proposed:

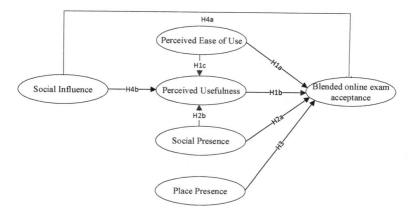

Fig. 11.1 Research model and hypotheses

H4a: Social influence positively affects learners' blended online examination acceptance.

H4b: Social influence positively affects perceived usefulness.

11.3 Research Methodology

11.3.1 Sample Plan and Data Collection

In this study, undergraduates in a university in Hubei Province were selected as the research object. A total of 992 questionnaires were distributed online. All the invalid questionnaires with the same answers, less than 60 s of filling time, and different forward questions and reverse questions were excluded. A total of 760 valid questionnaires were screened out, and the effective recovery rate reached 76.6%. In the basic information of the sample, male (41.97%) and female (58.03%); science (38.55%), liberal arts (21.84%) and engineering (39.61%); freshmen (34.21%), sophomores (29.08%), juniors (20.13%), and seniors (3.42%).

11.3.2 Questionnaire Development

The questionnaire is divided into two parts; the first part is personal basic information. The second part consists of 33 questions from 6 dimensions in the research model. Each dimension mainly refers to the existing maturity scales of perceived ease of use [24], perceived usefulness [24], social presence [25], place presence [26], social influence [27], and blended online examination [28] and modifies the relevant wording to meet the needs of this study.

11.4 Results

11.4.1 Analysis of the Measurement Model

We evaluated the measurement scales using the three criteria suggested by Fornell, namely (1) All indicator factor loadings (k) should be significant and exceed 0.5; (2) Construct reliabilities should exceed 0.8; (3) The average variance extracted (AVE) by each construct should exceed the variance due to measurement error for the construct (e.g., AVE should exceed 0.5). The Cronbach's alpha scores, shown in Table 11.1, indicated that each construct exhibited strong internal reliability, in which all the standard factor loading (k) values in confirmatory factor analysis of the measurement

Table 11.1 Construct reliability and convergent validity

Construct/indicator	Questionnaire items	Factor loading	Composite reliability (CR)	Average variance extracted	Cronbach's alpha
Perceived ease of use	PEOU1	0.77	0.90	0.68	0.91
	PEOU2	0.82			
	PEOU3	0.87			
	PEOU4	0.77			
Perceived usefulness	PU1	0.66	0.87	0.63	0.88
	PU2	0.68			
	PU3	0.72			
	PU4	0.58			
Social presence	SP1	0.73	0.89	0.68	0.88
	SP2	0.77			
	SP3	0.65			
	SP4	0.53			
Place presence	PP1	0.64	0.88	0.65	0.88
	PP2	0.84			
	PP3	0.70			
	PP5	0.72			
Social influence	SI1	0.72	0.90	0.70	0.88
	SI2	0.74			
	SI3	0.74			
	SI4	0.67			
Blended online examination acceptance	BTA2	0.53	0.77	0.64	0.90
	BTA4	0.85			

model exceeded 0.5 and were significant at $p = 0.001$. Besides, the composite reliabilities of constructs ranged from 0.81 to 0.93, and the AVE, ranging from 0.57 to 0.83, were greater than the variance due to measurement error. Therefore, all three conditions for convergent validity were met. According to Fornel, the correlations between items in any two constructs should be lower than the square root of the average variance shared by items within a construct. As shown in Table 11.2, the square root of the variance shared between the construct and its items was greater than the correlations between the construct and any other construct in the model, satisfying criteria for discriminant validity. All diagonal values exceeded the inter-construct correlations, and thus, the results confirmed that our instrument had satisfactory construct validity.

Table 11.2 Correlation matrices and discriminant validity

Construct	Perceived ease of use	Perceived usefulness	Social presence	Place presence	Social influence	Blended online examination acceptance
Perceived ease of use	0.68					
Perceived usefulness	0.66	0.63				
Social presence	0.75	0.57	0.68			
Place presence	0.70	0.57	0.80	0.65		
Social influence	0.80	0.61	0.77	0.77	0.70	
Blended online examination acceptance	0.75	0.46	0.75	0.71	0.81	0.64

Note All correlations significant at $p < 0.05$ except where noted. Diagonal elements are square roots of average variance extracted

11.4.2 Analysis of the Structural Model

We assessed the overall goodness-of-fit using the chi-square exam and other fit indices (namely GFI, AGFI, CFI, NFI, and RFI). The results show that χ^2/df is 3.841, GFI is 0.915, AGFI is 0.889, CFI is 0.955, NFI is 0.941, RMR is 0.032, and RMSEA is 0.061. Generally, fit statistics greater than or equal to 0.9 for GFI, NFI, and CFI and fit statistics greater than or equal to 0.8 for AGFI indicate a good model fit. Furthermore, χ^2/df is between 3 and 5, and RMSEA values ranging from 0.05 to 0.08 are acceptable. The fitting indexes show that our model obtains enough model fitting.

11.4.3 Hypotheses Testing

The path significance of each hypothesized association in the research model and variance explained (R^2 value) by each path was examined, and Table 11.3 shows the standardized path coefficients and path significances. The research result shows that variables influencing learners' blended online examination acceptance include perceived usefulness, perceived ease of use, social presence, and social influence. Place presence has no significant effect on blended online examination acceptance (H3 is supported). Among them, perceived ease of use, social presence, and social

Table 11.3 Summary of hypotheses exam

	Hypotheses	Direct effect	Indirect effect	Total effect	Result
H1a	Perceived ease of use -> Blended online examination acceptance	−0.18	0.07	−0.11**	No
H1b	Perceived usefulness -> Blended online examination acceptance	0.29		0.29**	Yes
H1c	Perceived ease of use -> Perceived usefulness	0.24		0.24**	Yes
H2a	Social presence -> Blended online examination acceptance	0.22	0.08	0.30**	Yes
H2b	Social presence -> Perceived usefulness	0.27		0.27**	Yes
H3	Place presence -> Blended online examination acceptance	0.08		0.08	No
H4a	Social influence -> Blended online examination acceptance	0.45	0.13	0.58**	Yes
H4b	Social influence -> Perceived usefulness	0.45		0.45**	Yes

Standardized estimates are shown. $*p < 0.05$. $**p < 0.01$. $***p < 0.001$

influence have a direct and positive impact on perceived usefulness, and the direct impact is 0.24, 0.27, and 0.45, respectively, (H1b, H2a, and H4a are supported). Perceived ease of use has a significant negative impact on learners' blended online examination acceptance (H1a is not supported). Perceived usefulness, social presence, and social influence have a significant positive impact on blended online examination acceptance (H1c, H2b, and H4b are supported). The direct impact effects are 0.29, 0.22, and 0.45, and the total impact effects are 0.29, 0.30 and 0.58, respectively. Besides, the result shows that both social presence and social influence have an indirect and significant positive impact on blended online examination acceptance through perceived usefulness, with an indirect effect of 0.08 and 0.13, respectively. Besides, it also shows the variance of the model, which explains a 72% change in perceived usefulness and a 72% change in blended online examination acceptance.

11.5 Discussion

Partly inconsistent with the findings of Joo et al. [9], perceived ease of use did not have a significant positive effect on blended online examination acceptance, but a significant negative effect. That is cognitive communication use in the online examination system awareness is not positive impact on learners' blended online examination acceptance. In the blended online examination, students' requirements for the system mainly involve the connection to the network, publication of exam rules and procedures, and the issuance and collection of examination papers. Blend online examination requires much less fluency, logic, agility, and complexity of running system than the ordinary online learning process in which learners search, acquisition and learning knowledge, as well as the improvement of personal information, activities, and interaction with teachers and peers in the ordinary online learning process. Therefore, in this context, the perceived ease of use has a small impact on the blended online examination acceptance. Combined with the limitation of the data range, it is concluded that the perceived ease of use has a significant negative impact on the acceptance of the blended online examination acceptance. Consistent with previous findings [6], perceived usefulness had a direct positive impact on the blended online examination acceptance. Online exam mode compared with the traditional exam model is more flexible and convenient, breaks through the limit of the space, can reduce the examinee to take an examination of additional consumption, such as time and energy. On this advantage, the online examination process and system can be improved to help improve the perceived usefulness of the candidates, to further increase the acceptance of the online examination.

The positive impact of social presence on blended online examination acceptance confirms the research of Palloff and Pratt [14]. At present, the traditional exam mode is still in the mainstream in the daily examination, except for some large-scale specific examination such as IELTS and TOEFL, computer level 2 exam, and Putonghua exam. The traditional examination mode is just in the transition stage to the online examination model, and learners are still more accustomed to the state of close connection with teachers and students in psychological distance and physical distance in the traditional examination mode. Therefore, social interaction and cognitive communication through video, voice, and other ways can enhance learners' social presence and improve their acceptance of the online exam.

The research results show that there is no direct and significant correlation between online learners' place presence and blended online examination acceptance. Place presence involves "immersive" learning in which the learner becomes an avatar in a simulated real environment, whereas in the context of the blended online examination, there is no need for rich and deep communication and interaction between learners and the environment, between learners and teachers, and between learners and learners themselves, and learners mainly spend their time and energy on completing the task of the examination. Therefore, there is no significant correlation between learner's place presence and blended online examination acceptance.

Besides, the research results also show that social influence has a positive impact on blended online examination acceptance and perceived usefulness, which confirms the previous research conclusions [22, 23]. The research results show that in this situation, social influence has the largest effect on the blended online examination acceptance, indicating that social recognition and support have a significant impact on the online examination acceptance of learners. The main reason is that the outbreak and spread of the epidemic situation in the background of this study disrupted the normal learning and living conditions of learners. The anxiety, instability, and insecurity of social emotions, as well as the carelessness of learners at home, make it difficult to concentrate on learning and achieve the teaching effect of the traditional classroom. The smooth development of online teaching is inseparable from the call of the government and the organization and planning of school teachers. The proposal of "suspended class, ongoing learning" proposed by the Ministry of Education is a good strategy to help learners complete their teaching tasks on time. Through the school, teachers, education departments at all levels of planning, and programs can enable learners to complete the task of learning. Before completing the examination, learners also realized the importance of online learning during this epidemic period through online learning for a period of time. Such perceived information will also generate and strengthen the perceived experience of usefulness for online learning and examination.

11.6 Implications

Based on current research findings, the following suggestions are proposed from the perspectives of the internal and external environment of online examination:

First, the online exam system and platform can be organized from the perspectives of functional design and development and exam service support. In the functional design and development phase, the goal is to match the learning assessment task and complete the process flow of the online examination model. In this phase, information technology plays a key role in the digitization of the test results to provide visualization and smart personalization requirements. It is also necessary for the educational administration department to perfect the online examination model and process to realize a smooth transition and transformation from the traditional examination model to online examination system. During the development process, various communication and interaction channels can create a trusting group atmosphere that can increase the social presence of learners.

Second, from the perspective of improving social influence, system planning should include the advocacy and suggestions of educational administration departments, the planning and supervision of educational departments at all levels, and the improvement of the online examination process and mechanism. In particular, system planning should strengthen the supervision of online examinations and ensure the validity and authenticity of the online examination results. These approaches will improve the acceptance of online examinations.

11.7 Limitation and Future Research

There are still some shortcomings in this study, and the main points are as follows:

First, the theoretical research model is not comprehensive. It is necessary to further study the influencing factors and correlation between learners' online examination acceptance in different periods and other dimensions in future studies. Second, with the promotion and popularization of online learning, in addition to primary and secondary school students, graduate students and above, and even social people outside the school are all beneficiaries of online examination. Therefore, comprehensive research on all segments of online learners is an important direction in the future.

References

1. Myyry, L., Joutsenvirta, T.: Open-book, open-web online examinations: developing examination practices to support university students' learning and self-efficacy. Active Learn. High. Educ. (2015)
2. Stowell, J.R., Bennett, D.: Effects of online testing on student exam performance and test anxiety. J. Educ. Comput. Res. 42(2), 161–171 (2010)
3. Davis, F.D., Bagozzi, R.P., Warshaw, P.R.: User acceptance of computer technology: a comparison of two theoretical models. Manage. Sci. 35(8), 982–1003 (1989)
4. Canziani, B., MacSween, S.: Consumer acceptance of voice-activated smart home devices for product information seeking and online ordering. Comput. Hum. Behav. 119, 15 (2021)
5. Al-Emran, M.: Evaluating the use of smartwatches for learning purposes through the integration of the technology acceptance model and task-technology fit. Int. J. Hum.-Comput. Interact. (2021)
6. Shin, N.: Online learner's 'flow' experience: an empirical study. Br. J. Edu. Technol. 37(5), 705–720 (2010)
7. Webster, J., Hackley, P.: Teaching effectiveness in technology-mediated distance learning. Acad. Manage. J. (1997)
8. Akour, H.: Determinants of Mobile Learning Acceptance: an Empirical Investigation in Higher Education. ProQuest LLC, Ann Arbor, MI. http://www.proquest.com/en-US/products/dissertations/individuals.shtml (2009)
9. Joo, Y.J., Lim, K.Y., Kim, E.K.: Online university students' satisfaction and persistence: examining perceived level of presence, usefulness and ease of use as predictors in a structural model. Comput. Educ. 57(2), 1654–1664 (2011)
10. Parker, E.B., Short, J., Williams, E., Christie, B.: The social psychology of telecommunication. Contemp. Sociol. 7(1), 32 (1978)
11. Akyol, Z., Garrison, D.R.: The development of a community of inquiry over time in an online course: understanding the progression and integration of social, cognitive and teaching presence. J. Asynchron. Learn. Netw. 12(3), 3–22 (2008)
12. Garrison, D.R., Anderson, T.: E-Learning in the 21st Century: A Framework for Research and Practice. Routledge (2003)
13. Gunawardena, C.N.: Social presence as a predictor of satisfaction within a computer-mediated conferencing environment. Am. J. Dist. Educ. 11(3) (1997)
14. Palloff, R.M., Pratt, K.: Building online learning communities. Technol. Pedagog. Educ. 14(3), 353–369 (2007)
15. Cobb, S.C.: Social presence, satisfaction, and perceived learning of RN-to-BSN students in web-based nursing courses. Nurs. Educ. Perspect. (Natl. League Nurs.) 32(2), 115–119 (2011)
16. Kang, M., Kang, J.: Investigating the relationships of social presence, satisfaction and learning achievement in the blended learning environment. EMBO J. 16(23), 6985–6995 (2008)
17. Ogonowski, A., Montandon, A., Botha, E., Reyneke, M.: Should new online stores invest in social presence elements? The effect of social presence on initial trust formation. J. Retail. Consum. Serv. 21(4), 482–491 (2014)
18. Biocca, F., Harms, C., Burgoon, J.: Toward a more robust theory and measure of social presence: review and suggested criteria. Presence 12(5), 456–480 (2003)
19. Sheridan, T.B.: Musings on telepresence and virtual presence. Presence Teleop. Virt. Environ. 1(1), 120–125 (1992)

20. Bulu, S.T.: Place presence, social presence, co-presence, and satisfaction in virtual worlds. Comput. Educ. **58**(1), 154–161 (2012)
21. Hossain, A., Quaresma, R., Rahman, H.: Investigating factors influencing the physicians' adoption of electronic health record (EHR) in healthcare system of Bangladesh: an empirical study. Int. J. Inf. Manage. **44**, 76–87 (2019)
22. Wu, B., Zhang, C.: Empirical study on continuance intentions towards E-learning 2.0 systems. Behav. Inf. Technol. **33**(10–12), 1027–1038 (2014)
23. Wu, B., Chen, X.: Continuance intention to use MOOCs: integrating the technology acceptance model (TAM) and task technology fit (TTF) model. Comput. Hum. Behav. **67**, 221–232 (2016)
24. Davis, F.D.: Perceived usefulness, perceived ease of use, and user acceptance of information technology. Manag. Inf. Syst. Q. **13**(3), 319–340 (1989)
25. Shea, P., Bidjerano, T.: Learning presence: towards a theory of self-efficacy, self-regulation, and the development of a communities of inquiry in online and blended learning environments. Comput. Educ. **55**(4), 1721–1731 (2010)
26. Slater, M., McCarthy, J., Maringelli, F.: The influence of body movement on subjective presence in virtual environments. Hum. Factors (2016)
27. Zainab, A.M., Kiran, K., Karim, N.H.A., Sukmawati, M.: UTAUT's performance consistency: empirical evidence from a library management system. Malays. J. Libr. Inf. Sci. **23**(1), 17–32 (2018)
28. Lin, W.-S., Wang, C.-H.: Antecedences to continued intentions of adopting e-learning system in blended learning instruction: a contingency framework based on models of information system success and task-technology fit. Comput. Educ. **58**(1), 88–99 (2012)

Chapter 12
In-Service Teachers' Attitude Towards Programming for All

Majid Rouhani and Victor Jørgensen

Abstract Coding instruction has increased widely throughout primary and secondary education in many countries, and educators are just beginning to understand the complexities of teaching students to code. A question raised is whether everyone should learn to code to be fully literate participants in our future society? What are the teachers' main arguments concerning the concept of programming for all? We investigate these questions from a teacher's perspective and aim to determine what attitudes teachers have towards programming for all. We gave teachers a task to describe their thoughts and perceptions on programming for all and collected data during a programming course for in-service teachers. This paper reports on preliminary findings. Although the vast majority in this study have positive attitudes towards programming, we can also see negative attitudes. These concerns are mainly related to lack of time. In-service teachers in this study believe that programming can be fun and engaging. They come with many arguments on reasons why they should include programming in school.

Keywords Online course · In-service teachers · Programming for all · Attitudes towards programming

12.1 Introduction

The trend to compose programming into the educational program will keep developing, grow, be driven through corporations like Code.org [1], Apple and Microsoft [2]. Youth are remixing, adjusting and making their applications, programming and computer games [2]. Programming is challenging to learn and teach and is a complicated subject that requires multilayer expertise [3, 4]. Teachers are often left to teach something that they do not feel comfortable with, considering that it may take about ten years to transform an amateur into a specialist developer [5]. On the one hand, we have the growing trend to include programming in schools and, on the other hand,

M. Rouhani (✉) · V. Jørgensen
Norwegian University of Science and Technology, 7491 Trondheim, Norway
e-mail: majid.rouhani@ntnu.no

© The Author(s), under exclusive license to Springer Nature Singapore Pte Ltd. 2022   149
Ó. Mealha et al. (eds.), *Ludic, Co-design and Tools Supporting Smart Learning Ecosystems and Smart Education*, Smart Innovation, Systems and Technologies 249,
https://doi.org/10.1007/978-981-16-3930-2_12

the complexity and challenges of learning and teaching programming that may put teachers in a demanding and challenging situation. In this study, we take a closer look at teachers' perspectives on 'programming for all' and ask the following research questions: Why should everyone learn to code to be fully literate participants in our future society? What are the in-service teachers' main arguments concerning the concept of programming for all? We investigate what attitudes in-service teachers have towards programming for all. This involves classifying the arguments as positive or negative attitudes. Many people need to be computer literate and sometimes use computational thinking (CT) to solve problems. Some of them will need to use programming at some point. This paper emphasizes that need and uses the words of teachers to make that case. The paper contributes to the body of knowledge around in-service teachers' training to meet the growing demand for teachers being prepared to teach programming in different subjects and school levels. In this study, we investigate in-service teacher's attitude participating in a university-level course. This group is diverse; some with no prior programming knowledge, some are already teaching programming and others intend to include programming as part of other disciplines.

We have organized this paper as follows: The next section presents a literature review on related work. In Sect. 12.3, we introduce the case and explain our methodology. Section 12.4 analyse the quantitative data around the main themes identified in the data set. Section 12.5 discusses the main findings in this study, and Sect. 12.6 present the conclusions and further work.

12.2   Literature Review

To improve the quality of lessons and teach effectively, Makhmudov et al. [6] suggest that teachers should have the ability to read and write simple computer programs; the ability to use computer programs and educative documentation; the ability to use computer terminology, particularly as it relates to hardware; the ability to recognize educational problems that can and cannot be solved using the computer; the ability to locate information on computing as it relates to education; and the ability to discuss the moral and human-impact issues as they relate to the societal use of computers as well as the educational use of computers. Recent efforts to revitalize the importance of CT aim to democratize computing knowledge as an essential body of knowledge that learners need to have to cope well with the twenty-first century [7]. CT is solving problems, designing systems and understanding human behaviour by drawing on the concepts fundamental to computer science [8]. Researchers and teachers foster CT through the use of programming [2]. Programming involves planning, testing, debugging and improving the source code [9]. Both concepts (programming and coding) are common in education, and there are advantages and disadvantages to using these words as synonyms. To investigate these research questions, we need to understand the teachers' perceptions of the underlying concepts such as computer literacy, CT, programming and coding, and motivational beliefs and attitudes.

12.2.1 Why Should Everyone Learn to Program?

Students need to be familiar with the fundamentals of programming a computer regardless of whether they intend to become computer programmers [10]. Coding as essential literacy is fundamental, and everyone should learn to code to be fully literate participants in our future society [11, 12]. Many researchers believe that engaging in programming tasks has unique potential in promoting higher-order thinking performance since it can enable students to obtain programming knowledge [13, 14]. Computing is defined as the act of using computers to gather, calculate, analyse, store, protect and present information faster and more accurately than can a human, and most often, less expensively [15]. Hence, programming is an essential part of computing. One argument for teaching everyone about computing is that we need more workers to program (jobs). Computing is part of students' lives; they live in a computational world. Computational thinking, computing literacy, productivity and broadening participation are other arguments used [14]. Including necessary coding in primary curricula provides teachers with an effective means of exercising their students' general and higher-order thinking skills [16]. To effectively promote students' learning performance in programming courses, it is necessary to include proper teaching strategies [17]. In this study, we take a closer look at teachers' perspectives on 'why everyone should learn to program'.

12.2.2 Motivational Beliefs and Attitude

Passionate views and perspectives towards a school subject are pertinent for learning and instructive achievement [18]. Malmi et al. [19] suggest that an improved understanding of the complex factors (such as the academic and social capital [20]) influencing students' internal factors, such as beliefs, emotions and attitudes, would help us develop more effective interventions to influence students' perceptions about learning to program and improve pass rates and learning results. Motivational beliefs in science have been shown to predict academic achievement reliably. These beliefs reflect the motivational value a person accredits to a subject or task and named value beliefs. Positive motivational opinions are associated with students' persistence in performing well even when their interest and intrinsic enjoyment decrease [21]. Besides attitude, self-efficacy perception also plays a role in the students' success in a computer programming course. Self-efficacy refers to the trust in the skill that the individual has to perform work [22]. In-service teachers often must teach programming as part of other subjects. Introducing programming in addition to the topics they already teach may complicate the situation for them. Therefore, motivational beliefs and positive attitudes are essential aspects of a programming course for this group of teachers. The question is whether teachers find it meaningful that everyone should learn programming or perceived as a disruption in teaching their subjects?

Meaningful information is bound to be remembered because it effectively inter-faces with propositional networks. Meaningfulness saves time and promotes learning [23]. In this context, there is a need to understand teachers' attitudes on teaching programming to all.

12.3 Case and Research Method

12.3.1 The Case: A University Level Programming Course for Teachers

Our university provides an online programming course for in-service teachers who teach programming courses or integrate programming into other disciplines. We have divided the program into two subjects, each of 7.5 credits and adapted the program for teachers who teach in grades 8–13 (secondary school). These courses are 'basic programming for teachers' and 'applied programming for teachers' and use Python as the programming language. After completing the program, we expect teachers to teach programming and use it in their subject area. Teachers participate with their school's support, committing to providing some free time to teachers to complete the course, though they continue their primary duties during the two semesters. The Ministry of Education partly covers the schools' additional costs. We conducted a pre-course survey to figure out what different levels and subjects the teachers are teaching. We received 186 responses (a rate of 93%). 78% are teaching at upper secondary school, and 20.4% teach at lower secondary school. More than 80% of the teachers are teaching STEM-related subjects. Others teach in different topics such as language, history, music, arts and crafts. 24% of participants have already taught programming in their schools. 61% of participants had some level of programming knowledge before starting the first course. The participants all come from the same country, but they represent various schools and districts. All schools follow the national curriculum. Participants have varying levels of teaching experience, ranging from new teachers to those who have been teaching for many years. The incentive to enrol in these subjects' courses varies as well. The majority claim they are inspired, although others are less certain that programming should be included in their classes. We have not addressed demographic information in this study as we aim to include a large group with different backgrounds, needs, desires and motivations. The first subject curriculum covers basic programming knowledge, while the second subject covers different aspects of how to teach programming. When teachers start the second course, they start working on definitions of some important concepts related to computing education, e.g. 'why should everyone learn to code' and 'what are the expected challenges'. Teachers use approximately three weeks on this part of the course. Before continuing to other topics, they deliver an assignment to reflect on their attitude towards programming for all and teaching programming. We informed teachers that anonymized data will be used in research and had the option of not being included. The question we asked participants was: 'why should everyone learn to code?'

12.3.2 Research Method

Data Collection. We collected data from participants' reflection notes from the 2019 and 2020 cohorts to answer our research questions. We analysed a total of 328 reflection notes. For 2019, there were 22 out of 70 notes available for data analysis. For 2020, 169 out of 170 was used at the start and 137 at the end of the course. The main themes identified and discussed in this paper are: 'arguments for why everyone should learn to program, or why we should teach programming in school', and 'the attitude of teachers towards programming for all'. The research team processed the document's qualitative data in several steps:

- Documents were anonymized outside of the research team. Thus, a collection of reflection notes for the period 2019 and 2020 was available to the research team for further analysis.
- Researchers have an in-depth knowledge of the course. They are involved in developing the program and worked as teacher assistants. Hence, well known to the reflection notes in this study.
- The analysis started with themes already identified during the analysis of a similar data set [24]. The previous study, involving a cohort of teachers attending the same programming course, explored the impact of the training program on teachers' self-efficacy. Though the research questions addressed in the two studies are different, they share the overall aim of understanding challenges connected to programming and implications for professional development.
- To limit the thread to validity connected with a single coder, we added an initial phase for quality assurance. Few reflection notes were first coded together and discussed to familiarize the coder with the existing codebook. The coder then went ahead, coding additional notes.
- Researchers coded few documents independently to check inter-coder reliability. The method 'Percent Agreement for Two Raters' was used. We manually compared the codes and gave the score of 1 to those that are the same and 0 to those that are different. The number of 1s divided by the total of codes compared will be inter-rater reliability. The results from the inter-coder reliability check (two times during the study) were considered sufficient ($\approx 88\%$ in average).

Limitations and Threats. By analysing teacher's reflection notes, we looked for their perspectives on 'programming for all'. To ensure high validity, we used a significant sample size (328 reflection notes for two years) for this study, and the teacher's reflected without being directed by any questionnaire. Participants differ in terms of the subjects they teach, school type and motivational factors. We used a pre-existing codebook from a previous study involving a cohort of teachers attending the same programming course, which explored the impact of the training program on teachers' self-efficacy. Based on the research questions, the research group added new codes. All researchers in this study have an in-depth knowledge of the course and well known to the reflection notes being analysed. In-service teachers' attitudes and arguments may have been affected by the literature study in the course. Most participants are already aware of the benefits and importance of learning programming. Results are mainly from one course on programming for teachers and their experiences, and it might not be easy to generalize the results. Researchers need to investigate how programming is taught to this group at other institutions and countries compared to this approach.

12.4   Analysis

The summary of the data gathered is shown in Table 12.1. Reflection notes collected at the beginning of the course are related to the question: 'why should everyone learn to code?' This is the case both for 2019 and 2020. At the end of the course, many participants reflected on these questions without explicitly asking them to do so. A teacher may have several complementary or opposing reasoning. In this analysis, we only look at the number of arguments and not which teacher these belong to. In the following, we will give a brief explanation for each category and some of the results.

Table 12.1   Overview of themes—at the beginning and end of the course

	Year	Theme	# teachers	# code references
Start	2019	Arguments for learning to program	19	25
Start	2019	Arguments for not learning to program	4	12
Start	2019	Critical towards programming in school	2	5
Start	2019	Positive towards programming in school	16	31
Start	2020	Arguments for learning to program	214	252
Start	2020	Arguments for not learning to program	5	5
Start	2020	Critical towards programming in school	14	25
Start	2020	Positive towards programming in school	100	133
End	2020	Critical towards programming in school	6	7
End	2020	Positive towards programming in school	64	92

12.4.1 Teachers Attitudes and Arguments Towards Programming for All

Arguments. We used this theme to identify teachers' perceptions of why everyone should learn to program. We were looking to see to what extent teachers feel that these arguments align with their expectations and practices. We would also identify which reasons they categorize as most important and whether they possibly have other ideas that they believe are important for everyone to learn or not learn to program. A brief explanation of the codes with the most frequency is presented in Table 12.2.

The Attitude of Teachers. We used the theme' attitude of teachers to classify their views on programming for all. The category consists of two codes: 'Positive towards programming in school' and 'Critical towards programming in school'. For 2020 data collected at the end of the course, the code 'Positive towards using coding interdisciplinary' has also been used. A summary of themes and the number of participants is found in Table 12.1.

Positive Towards Programming in School. One hundred sixteen teachers have positive attitudes towards programming. Table 12.3 shows an overview of the codes with the most frequency.

Critical Towards Programming in School. Twenty-two is crucial or questionable against programming for all. These concerns are related to lack of time, colleges that are skeptical of introducing programming, lack of motivation, not seeing the benefits, etc. As part of other subjects such as mathematics, programming will cause less time to learn mathematics. One teacher write: 'Mathematics has been given responsibility for the teaching of programming in schools, and several critical voices have been afraid that the programming part may contribute to less time for the subject of mathematics'.

12.4.2 Arguments for Not Learning to Program

Nine teachers argue that programming is not relevant in some subject areas like health care, construction and electrician. Here are some examples: 'If we think of a student in secondary school, for example, in construction, I see little opportunity to apply programming in that area'. Another teacher writes: 'But personally, it is a bit difficult to see the usefulness of programming when working as an electrician. In my time out at work, I have never been involved in programming, nor have I heard anyone in my company doing this. Feel it is much more relevant for pupils in automation and computer electronics'. Another argument that speaks against programming for all is the high costs of teachers' education and device acquisition. A teacher writes:

Table 12.2 Teachers' arguments towards programming for all

Code	Description	Teacher quotes
A tool for better understanding the subjects	Sixty-eight teachers seem to be concerned with using programming as a tool in the subjects they teach. In this way, they may increase learning outcomes in subjects where programming is applied. Our data show that teachers are most interested in being able to use programming in other subjects	'In physics, we work a lot with formulas and calculations. Here the pupils can program solution formulas and let the computer carry out the calculations in different tasks. Through this, the pupils will learn the formulas better, and they will also have to reflect on what conditions the formulas apply to'
Abstract thinking capabilities	Forty-three argue that learning to think abstractly is essential for developing problem-solving skills. Pupils become more adept at thinking about a solution to a problem without trying it out	'In secondary school age, many pupils can think abstractly; it is an excellent time to work with algorithmic thinking to achieve algorithmic competence in the long run'
Recruiting	Thirty-six teachers argue that programming is the most crucial job skill in the future. Today's and the future's society requires workers who can program. Introducing programming in primary school will increase interest and ensure that more people choose fields of education in programming	'I teach in a lower secondary school. Many of our pupils have begun to think about their future and what they want to be. Many of our pupils have already become interested in IT and programming and want to learn more about this. We have offered programming as an elective for three years, and 15–20 pupils out of 120 have chosen this every year. This shows that there is an interest in learning this'
Interesting, fun and engaging	Twenty-eight teachers believe that programming can be fun and can create enthusiasm in subjects like robotics, control of electrical systems, etc.	'Programming can "trigger"/create enthusiasm for subjects that have not previously been so accessible in schools, such as robotics, control of electrical systems, Lego systems, and development of innovative solutions/apps and not least, it can create curiosity and creativity'

Table 12.3 Attitudes of teachers: positive towards programming

Code	Description	Teacher quotes
Joy and engagement	Programming can evoke joy, commitment, curiosity and be fun	'What gives increased interest and joy? Programming can increase pleasure. Succeeding at some basics, like coding a program that calculates interest, that you can be allowed to spend on trials. I think it creates joy, learning, and curiosity'
Future opportunities	Computing education, including programming, is essential to all, and everyone should get equal opportunity to succeed in the future job market	'It is essential to teach computer knowledge and technology to everyone to give everyone an equal opportunity to succeed in the job market. Minorities and other vulnerable groups will not be able to acquire this knowledge independently. Therefore, it must enter the school—both in terms of job opportunities and understanding society and one's own life'
Time-consuming process	Learning programming might be a time-consuming process. It may, therefore, be advantageous to start this process early. This subject area is large and demanding, which requires learning over many years	"That I make sure that pupils at least get a taste of what they are surrounded by continuously, I consider very natural and important, regardless of whether you end up in this industry or not. It is natural to learn languages such as Norwegian, English, German, Spanish, etc. For many, learning different programming languages will be just as essential"
Contribution to better understanding of other subjects	Many teachers believe that programming help pupils understand other topics better	'Programming may be a useful tool for solving various problems and challenges that we face and better understand other subjects such as STEM-related subjects'

'Arguments that can be used against learning all programming are high costs associated with teachers' hardware and education. Pupils who do not like problem-solving can have an additional discouraged school day'.

Some of the teacher responses seem to contain misconceptions, e.g. the implication that programming is counter to students' wanting to 'work with people and relationships' and give 'cheaper' to learn to program because one senior teacher is cheaper than multiple junior tutors. Student misconceptions and judgments of

Table 12.4 Reference to arguments for learning to program form the textbook

Code	# references
Broadening participation	239
Computational literacy	299
Computational thinking	194
Jobs	320
Learn about world	290
Productivity	139

programming and firmly related subjects have been investigated for quite a while [25–27].

12.4.3   Arguments for Learning to Program from the Literature

In this study, we are interested in what arguments in-service teachers believe are most important to them. Views referenced by teachers without further reflection are categorized under the theme 'Arguments from the literature' and omitted in this study. The number of teachers that used these arguments is 190, with 1706 codes identified which are overlooked and not further analysed. Table 12.4 shows the code frequency.

12.5   Discussions

Our study results indicate that in-service teachers are positive towards programming for all, even though this is challenging and difficult to achieve [3, 4]. Teachers in this context are most concerned with using programming as a tool to increase the pupil's understanding of multidisciplinary subjects. Many in-service teachers are short on time, and therefore the application of programming in their discipline may be a motivating factor. In this way, they may increase learning outcomes in subjects where programming is applied [13, 14]. Abstract thinking capabilities is another area that concerns highly in-service teachers in this study. Computational thinking can be considered a problem-solving toolset that applies to computing principles such as abstraction [7, 28] and has been offered as a cross-disciplinary set of mental skills [29]. Many in-service teachers agree that this is an essential component of developing problem-solving skills in interdisciplinary subjects and promoted higher-order thinking [13, 14]. Many in-service teachers argue that programming is one of the most crucial job skills in the future, and at the same time, a time-consuming process; therefore, pupils must start learning to program early. Several teachers express that programming may even out differences in the population. Everyone will have the same opportunity, regardless of social background and gender. In the literature, we find that computing education, including programming, is essential to all, and everyone should get an equal opportunity to succeed in the future job market [14]. Technology discoveries will create millions of new jobs in programming and related areas. It will be challenging to find talents with the required education and skills for these occupations [30]. Recruitment seems to occupy in-service teachers significantly. However, they find some of these arguments in the textbook [14], which may have influenced their perceptions after becoming more proven on this idea. Teachers mention few other views, such as 'programming may bring innovation' and 'develop pupil's creativity skills'. There are, however, relatively few teachers who are engrossed in these. In-service teachers also express scepticism about introducing programming as part of other subjects (such as mathematics), which may go beyond their allocated time.

The results of this study indicate that teachers believe programming can be engaging and fun [2]. Although the vast majority in this study have positive attitudes towards programming, we can also see negative attitudes. These concerns are mainly related to lack of time. Humble and Mozelius [31] also report this issue in their article, where they discuss obstacles and opportunities in integrating programming in the K-12 setting. Having enough time to teach students how to program and using it as a tool for developing other skills is highly important [32]. Introducing programming as part of other subjects might complicate teachers' situation and negatively affect the learning outcomes. The schools receive some government financial support for sending in-service teachers to these courses. Few reports that their motivation for attending the program has been to achieve credits (ECTS) and not being motivated to learn to program and believe that everyone does not need to know how to program. We find some of the same arguments among other critics[1] of 'programming for all'[2] (future jobs, computational thinking, etc.) [33]. Our results show 25 references to being critical towards programming at the beginning of the course (see Table 12.1), while we found only seven references for the same theme at the end of the course in 2020. There is no link between these results; those seven references may be from the same teachers or different ones. This decline in the number of in-service teachers being critical to 'programming for all' may indicate a change in their attitudes during the course. However, we have not focused on studying whether teachers in this course change attitudes along the way.

12.6 Conclusion and Future Work

The preliminary results of a study on how in-service teachers conceive and interpret programming for all and why programming should (or should not) be included in

---

[1] https://techcrunch.com/2016/05/10/please-dont-learn-to-code/.

[2] https://www.wise-qatar.org/coding-cognitive-abilities-michael-trucano/.

school curricula were summarized in this paper. The paper investigates the value of computational thinking, coding and programming in school and in-service teachers' reactions to the prospect of programming becoming a full autonomous subject matter. Most of the in-service teachers in this study believe that programming can be fun and engaging. They come up with many arguments and reasoning for why they should include programming in school. Considering that this type of training requires a substantial investment, both at the individual and the school level, we claim it is essential to investigate teacher's attitudes towards programming for all.

This study suggests that in-service teachers may be concerned with using programming as a tool in the subjects they teach. In this way, they may increase learning outcomes in subjects where programming is applied. Simultaneously, they express scepticism about introducing programming as part of other subject areas, which may go beyond the allocated time available. Although the vast majority in this study have positive attitudes towards programming, we can also see negative attitudes. These concerns are mainly related to a lack of time and disagree that everyone must learn to program. We are fully aware that in-service teachers' attitudes and arguments for why everyone should learn programming may have been affected by the literature study (textbook) in the course. Another limitation of this study is that most participants are already aware of the benefits and importance of learning programming (compared to the ones less convinced of its value and hence would not be in this course). Participants differ in terms of the subjects they teach, school type and motivational factors. Therefore, the study covers different perspectives. The main contribution of this paper is that our data indicates a clear positive evolution of the in-service teacher's perception towards the dissemination of coding and programming in school curricula.

Although the results are promising, they are mainly from one course on programming for teachers and their experiences, and it might not be easy to generalize the results. Researchers need to investigate further how programming is taught to this group at other institutions compared to this approach. Studies across countries would be relevant to understanding how the educational system and other programming courses impact teachers' attitudes. Many of the teachers participating in this course had not yet taught programming in their classes. So, they were not speaking from that experience. A follow-up study might be relevant after they have included programming in their courses for a couple of years.

References

1. Kalelioğlu, F.: A new way of teaching programming skills to K-12 students: Code. org. Comput. Hum. Behav. **52**, 200–210 (2015)
2. Cooke, L., et al.: Can everyone code?: preparing teachers to teach computer languages as a literacy. In: Participatory Literacy Practices for P-12 Classrooms in the Digital Age, pp. 163–183. IGI Global (2020)
3. Cheah, C.S.: Factors contributing to the difficulties in teaching and learning of computer programming: a literature review. Contemp. Educ. Technol. **12**, ep272 (2020)

4. Jiau, H.C., Chen, J.C., Ssu, K.-F.: Enhancing self-motivation in learning programming using game-based simulation and metrics. IEEE Trans. Educ. **52**, 555–562 (2009)
5. Winslow, L.E.: Programming pedagogy—a psychological overview. ACM SIGCSE Bull. **28**, 17–22 (1996)
6. Makhmudov, K., Shorakhmetov, S., Murodkosimov, A.: Computer literacy is a tool to the system of innovative cluster of pedagogical education. Eur. J. Res. Reflect. Educ. Sci. **8** (2020)
7. Angeli, C., Giannakos, M.: Computational thinking education: issues and challenges (2020)
8. Wing, J.M.: Computational Thinking (2008)
9. Blackwell, A.F.: What is programming? In: PPIG, p. 20 (2002)
10. Popyack, J.L., Herrmann, N.: Why everyone should know how to program a computer. In: IFIP World Conference on Computers in Education, pp. 603–612 (1995)
11. Hutchison, A., Nadolny, L., Estapa, A.: Using coding apps to support literacy instruction and develop coding literacy. Read. Teach. **69**(5), 493–503 (2016)
12. Martin, C.: Coding as literacy. In: The international encyclopedia of media literacy, pp. 1–8 (2019)
13. Wang, X.-M., et al.: Enhancing students' computer programming performances, critical thinking awareness and attitudes towards programming: an online peer-assessment attempt. J. Educ. Technol. Soc. **20**, 58–68 (2017)
14. Guzdial, M.: Learner-centered design of computing education: research on computing for everyone, p. 147
15. Cortada, J.W.: What is computing? In: Living with Computers, pp. 1–4. Springer (2020)
16. Falloon, G.: An analysis of young students' thinking when completing basic coding tasks using Scratch Jnr. On the iPad. J. Comput. Assist. Learn. **32**, pp. 576–593 (2016)
17. Adu-Manu, K., Arthur, J., Amoako, P.: Causes of failure of students in computer programming courses: the teacher learner perspective. Int. J. Comput. Appl. **77**, 27–32 (2013)
18. Leifheit, L., et al.: Development of a questionnaire on self-concept, motivational beliefs, and attitude towards programming. In: Proceedings of the 14th Workshop in Primary and Secondary Computing Education, pp. 1–9 (2019)
19. Malmi, L., et al.: Theories and models of emotions, attitudes, and self-efficacy in the context of programming education. In: Proceedings of the 2020 ACM Conference on International Computing Education Research (2020)
20. Bourdieu, P.: The forms of capital (1986). In: Cultural Theory: An Anthology, vol. 1, pp. 81–93 (2011)
21. Oliver, J.S., Simpson, R.D.: Influences of attitude toward science, achievement motivation, and science self concept on achievement in science: a longitudinal study. Sci. Educ. **72**, 143–155 (1988)
22. Horzum, M.B., Cakir, O.: The validity and reliability study of the Turkish version of the online technologies self-efficacy scale. Educ. Sci. Theory Pract. **9**, 1343–1356 (2009)
23. Schunk, D.H.: Learning Theories: An Educational Perspective, 6th edn. (2012)
24. Thorsnes, J., Rouhani, M., Divitini, M.: In-Service Teacher Training and Self-efficacy
25. Mayer, R.E.: The psychology of how novices learn computer programming. ACM Comput. Surv. (CSUR) **13**, 121–141 (1981)
26. Bayman, P., Mayer, R.E.: A diagnosis of beginning programmers' misconceptions of BASIC programming statements. Commun. ACM **26**, 677–679 (1983)
27. Clancy, M.: Misconceptions and attitudes that interfere with learning to program. Comput. Sci. Educ. 85–100 (2004)
28. Selby, C.C.: Relationships: computational thinking, pedagogy of programming, and Bloom's taxonomy. In: Proceedings of the Workshop in Primary and Secondary Computing Education, pp. 80–87 (2015)
29. Yadav, A., et al.: Computational thinking in teacher education. In: Emerging Research, Practice, and Policy on Computational Thinking, pp. 205–220. Springer (2017)
30. Gordon, E.E.: Future Jobs: Solving the Employment and Skills Crisis (2013)
31. Humble, N., Mozelius, P.: Teacher perception of obstacles and opportunities in the integration of programming in K-12 settings. In: Proceedings of EDULEARN (2019)

32. Jawawi, D.N.A., et al.: Introducing computer programming to secondary school students using mobile robots. In: 2015 10th Asian Control Conference (ASCC), pp. 1–6 (2015)
33. Tellidis, E.G.: Should we teach our students computer science? In: Education and Technology: Manitoba Makers and Coders, p. 147 (2017)

Chapter 13
Robots as My Future Colleagues: Changing Attitudes Toward Collaborative Robots by Means of Experience-Based Workshops

Janika Leoste, Tõnu Viik, José San Martín López, Mihkel Kangur, Veiko Vunder, Yoan Mollard, Tiia Õun, Henri Tammo, and Kristian Paekivi

Abstract Artificial intelligence-driven robots are increasingly being introduced in various workplaces. Research implies that people's negative attitudes toward intelligent and collaborative robots might hinder their willingness to use them. We propose that interactive educational activities such as specialized workshops help people to overcome such negative attitudes. We designed a two-day workshop that introduced two quasi-industrial robots (Poppy Ergo Jr and ClearBot) to 16 university students. Students' attitudes were qualitatively measured before and after the workshop. The results imply that the workshop helped students to increase their understanding of the nature of the intelligent collaborative robots. More precisely, robots became to be seen as empowering tools, rather than friends or enemies. Interestingly, there were significant gender differences, as the female participants had a greater tendency to view robots as animated objects. We concluded that specialized workshops effectively lead participants to become aware of various promising opportunities for their robotic co-workers in the possible future.

Keywords Attitudes toward robots · Human–robot collaboration · Artificial intelligence · Machine learning

The original version of this chapter was revised: This chapter has been changed as Open Access licensed under the terms of the Creative Commons Attribution 4.0. The correction to this chapter can be found at
https://doi.org/10.1007/978-981-16-3930-2_18

J. Leoste (✉) · T. Viik · M. Kangur · T. Õun · H. Tammo · K. Paekivi
Tallinn University, 10120 Tallinn, Estonia
e-mail: leoste@tlu.ee

J. S. M. López
Universidad Rey Juan Carlos, Calle Tulipán, s/n, 28933 Móstoles, Madrid, Spain

V. Vunder
Institute of Technology, University of Tartu, Tartu, Estonia

Y. Mollard
Poppy Station, 75007 Paris, France

© The Author(s) 2022, corrected publication 2022
Ó. Mealha et al. (eds.), *Ludic, Co-design and Tools Supporting Smart Learning Ecosystems and Smart Education*, Smart Innovation, Systems and Technologies 249,
https://doi.org/10.1007/978-981-16-3930-2_13

13.1 Introduction

How will we feed an ever-growing population, provide clean water, generate renewable energy, prevent and cure disease and slow down global climate change?

I hope that science and technology will provide the answers to these questions, but it will take people, human beings with knowledge and understanding, to implement the solution.

Hawking [1]

Frontier technologies are promising for meeting the challenges of the world with limited resources, making sustainable development a reality, while protecting the planet. One of these frontier technologies is robotics, combined with machine learning and artificial intelligence [2]. Such advanced robotics could help meeting various goals, from offering personalized health services to offering affordable products and services, while saving energy and material resources [2–4]. The number of robots around us is already constantly increasing [5]. It is believed that this tendency will lead to robots becoming a common phenomenon in people's everyday lives in many countries worldwide [6]. However, the presence of the robotics technology may become a source of great discomfort for many people for various reasons. Frequently, questions about the potential challenges or threats to humanity are raised [7, 8], but even more often people express their fear of losing jobs or weakening of interpersonal relationships; i.e., they are afraid of becoming marginalized [6, 9]. The likelihood of negative attitudes grows together with the prevalence of robots in people's everyday lives [6], obstructing the effective use of this promising technology.

Literature suggests that, similarly to any other type of objects, attitudes toward robots are based on people's overall evaluations [8]. Although the idea of robots is generally perceived rather positively, people's attitudes in more specific contexts are determined by the usage scenarios of robots. For example, the studies about a robotic vacuum cleaner Roomba imply that its users find it as a likable, if not even pet-like object [10]. In some other contexts, robots could provide a sense of security. However, robots are also described as "uncanny" or "eerie" when their intended use includes close interaction with children or elderly in education and elder care, or when they are used for leisure [8, 11]. On the other hand, the differences between interacting with one or more robots must also be taken into account. Apart from their role in the interaction, groups of robots are being more accepted in tasks as functional robots rather than as social robots. On the contrary, an individual social robot is more accepted than a functional one [12]. In addition, in many working environments workers relate to robots as social entities with whom they interact. In these interactions, workers rely on cues in order to understand and predict robots' actions—a precondition for workers to feel safe when near robots. The experience from these interactions also determines workers attitudes toward robots [10, 13]. It is suggested that people's attitudes toward robots depend on their awareness about the possibilities and features of modern robotics [14]. Previous experience with robots tends to increase positive attitudes toward them and leads to higher robot acceptance by people [15, 16].

Attitudes toward robots can be used as precursors for predicting people's willingness to use robotics technology (see also [17–19]), suggesting that by changing people's attitudes toward robots it is possible to reduce their robot anxiety [11]. One of the tools for introducing systemic change in people's attitudes and behaviors is using workshops. Workshop is a learning method that through a short-term learning experience encourages active learning [20]. Through active learning, it is also good to support the development of people's key competences such as adaptability, creativity, curiosity, and open-mindedness, which will help them to cope in the labor market of the future [21]. The workshop format implies an active engagement of a relatively small group of students interacting among themselves, and with the teachers. Workshops should follow a well-defined structure, including an introduction, interactive learning activities that are matched to set objectives, and a conclusion [22]. The overall design of a workshop should meet the needs of adult learners' limited attention span of up to 20 min and provide learners with interactive learning and material, that is, practical, contextual, and applicable [23].

Based on these notions, we propose a two-day experience-based workshop on two specific robots (that serve as models of real-life industrial robots) designed as a familiarization process, aiming to grow people's positive attitudes toward robots. Our hypothesis is that people who participate in a workshop involving immediate first-hand engagement with collaborative robots' prototypes will develop a more adequate understanding of robots and their features, which in turn triggers the attitude change toward robots in general. In addition, we use the elements of user innovation process [24] by asking the participants to reflect on their experience about the workshop.

The focus of this paper is guided by the following research questions:

1. Do participants' intentions about collaboration with robots in the future change after the experience-based workshop?
2. What are the participants' suggestions on improvement of the experience workshop design?

13.2 Materials

13.2.1 Robots and Artificial Intelligence

Robot is a machine that is capable of doing something automatically. Robots are usually designed for some specific tasks that determine its appearance, smartness level, and other features. Any robot can be described by the following four characteristics [14]:

- How they relate to their operating environment (on the ground, in the air, underwater, in space, inside the human body);
- how they interact and collaborate with users (programmed, tele-operated, supervised, collaborative, or autonomous);

- the design of their physical appearance (robot arms, robot platforms, exoskeletal robots, metamorphic robots, nano- and micro-robots, humanoid);
- their primary function (assembly of parts, surface processing, interaction, exploration, transporting, inspection, grasping, or manipulation).

Robots are increasingly driven by artificial intelligence (AI), which allows robots to perform tasks that normally would require human intelligence [25]—for example, to drive a car or paint a picture. AI can be "weak" (the robot simulates human intelligence) or "strong" (the robot is able to optimize its behavior based on its former behavior and its experience) [26]. While AI gives rise to various questions, from machine ethics to robot rights, its use allows robots to function more efficiently and can give them pet-like or human-like features. A component of AI is machine learning (ML)—computer algorithms that improve automatically through experience [27]. A ML-driven robot uses sample data to create a model of something (e.g., a map of its surroundings). Today, ML is used in many domains, from banking to quantum chemistry, allowing computers to find and examine connections and relationships between numerous factors and to make predictions or decisions that would be challenging for humans.

An increasingly important type of robot is *cobot*—a collaborative robot that is designed to interact, collaborate, and work with a human operator [10, 28]. The connection between a cobot and its human operator is created via virtual surfaces or other haptic effects within a workspace shared by a human and cobot. Cobots can enhance human abilities—for example, exoskeletons provide mobility for disabled persons or greater strength for workers [29, 30]. However, more and more frequently, process automation has filled our environment with cobots that must meet the following requirements (e.g., for collaborative industrial tasks): (1) ensure that the operations of a cobot can be dynamically modified by an operator; and (2) produce flexible and adaptive behaviors with the human beings who handle them [31]. The robots that we are using as the basis for this study are simplified models of AI-driven cobots.

13.2.2   ClearBot and Poppy Ergo Jr

In this study, we use two educational robots, ClearBot, and Poppy Ergo Jr to conduct two workshops in order to familiarize participants with the concepts of machine learning and artificial intelligence.

ClearBot[1] (Fig. 13.1) is used for teaching the principles of wheeled mobile robotics and computer vision. Its dimensions are about 30 × 30 × 30 cm (depending on installed hardware). It features Intel Core i5 (7th Gen) 7260U CPU, Intel Iris Plus Graphics 640 GPU, Intel RealSense D435 3D Depth Camera, three omnidirectional wheels with separate motors, and a transparent design. ClearBot programming is

---

[1] https://clearbot.eu/technology/.

Fig. 13.1 ClearBot (left) and Poppy Ergo Jr (right)

done via robot operating system (ROS)—a set of software libraries and tools for building robot applications.

Poppy Ergo Jr[2] (Fig. 13.1) is a robotic arm that was designed for classroom use with the goal of introducing students to subjects like programming, geometry, robotics, engineering, and design. Poppy Ergo Jr uses a small single-board computer Raspberry Pi as its control module, and six servomotors for life-like movements. This robot can be programmed in a visual programming language Snap (a variant of Scratch), Python, or ROS.

13.3 Method

We designed behaviors for two affordable educational robotics platforms (Poppy Ergo Jr and ClearBot) with the aim of using these in a two-day experience workshop "My Future Colleague Robot" (Fig. 13.2). In the context of this workshop, the Poppy Ergo Jr and ClearBot robots can be considered as simplified models of industrial robots. The goal of the workshop is to introduce AI-driven robots to non-ICT field teachers, university students, and pupils, to encourage discussion about robots as their future colleagues, and to initiate a positive change in the participants' attitudes toward them.

Our sample consisted of interdisciplinary BA and MA level higher education students ($n = 16$, 9 men, 7 women, minimum age 20, maximum age 50, average age 33), from the following fields: organizational behavior, IT management, andragogy, class teacher, special educator advisor, early childhood education, math, educational sciences, informatics, psychology, nature sciences. All these students were registered to six credit LIFE course (a course focused on collaborative real-life challenge solving) "My Future Colleague Robot." One of their course assignments was to evaluate appropriateness of ready-made robotic behaviors for two robotic platforms.

---

[2] https://www.poppy-project.org/en/robots/poppy-ergo-jr/.

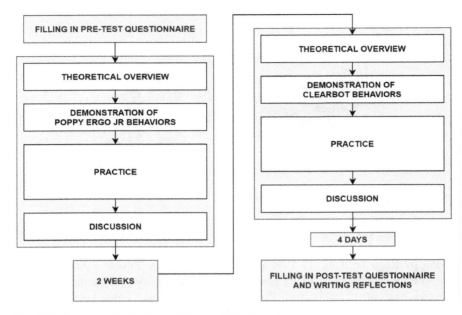

Fig. 13.2 Structure of "My Future Colleague Robot" workshop

We used an online questionnaire and participant reflections for **data collection**. The questionnaire recorded participants' background information and their answers in open text format to two questions:

1. Think about how your professional life will look like in 10 years from now. Do you see intelligent robots playing a role in it? What could these roles be like? While it is understandable that you do not know precisely what you will be doing after 10 years, please imagine how your future work-life could look like. Will the robots be your friends or foes, colleagues or tools?
2. Think about your personal life after 10 years from now. Do you see intelligent robots filling some roles in it? What could these roles be like? While it is understandable that you cannot know precisely how your life will turn out in the future, please imagine how your future private life could look like. Will robots be your friends or foes, family members or slaves—or just tools?

The participants were asked to fill in the questionnaire before and after the two-day experience workshop where ready-made robotic behaviors were demonstrated to participants. The posttest questionnaire had to be filled in within four days after the second workshop took place. In addition, the participants were asked to write a reflection on their feelings and attitudes; they had during the workshop. They were also asked to provide a few suggestions on how to improve the experience workshop so that it could help the members of different target groups (scholars, higher education, and general education students) to enhance their understanding and perception about cobots.

We used quantitative content analysis [32] with closed coding for analyzing the answers to the questionnaires, and interpretative phenomenological analysis [33] for analyzing the reflections.

The contents of the first day of the experience-based workshop included the following:

- Participants filled in the pretest questionnaire before the experience workshop started. The questionnaire was formed in Google Form and participants used their own smart devices (15 min).
- Researchers presented definitions for robots, AI and ML and gave an overview of a number of studies on people's attitudes about robots (15 min).
- Researchers demonstrated two Poppy Ergo Jr robot behaviors. The "Learn and Show" behavior is about teaching the robotic arm some certain movements by moving it first as desired. The machine stores the movement and repeats it on command. Similar robots in the industry can be reused at various tasks (as compared to specialized robots that are fit for only specific tasks). With the "Sorter" behavior, the robot is taught to react differently to two different colors, allowing the robot to perform different tasks on factory conveyor belts (e.g., sorting items) (15 min).
- Participants tried these behaviors out by themselves and discussed with researchers on the applicability of these behaviors in a real-life context (30 min).
- Poppy Ergo Jr robotics platform developers presented it in greater depth. Participants were told the story of this robot and its other versions. The current usage cases by different user profiles in research, art, and education were presented. The development process that led the engineering team to design Poppy Ergo Jr with its current mechatronics, software, pedagogical activities, and open-source community was explained (15 min).

The second day of the experience-based workshop was conducted after two weeks with the following contents:

- Researchers introduced the concept of sustainability in the context of robotics. The lecture focused on global challenges related to technological development, and people's perceptions on the future perspectives of solving the current environmental issues with frontier technologies, focusing on how and for what purposes the technologies can be used as improperly applied technologies can exacerbate problems [34] (15 min).
- Researchers demonstrated three ClearBot robot behaviors with the purpose of helping participants to build better understanding about collaborative robots, and to demonstrate how these robots can be used outside the industrial context, for example, as companions or helpers. The goal of the "Teleoperation—delivering the water bottle to a COVID patient" behavior was to show that a person does not always have to move to take something. Via teleoperation, a person can avoid challenging/dangerous areas or close contact with others, etc. A robot acts as a facilitator in this scenario. The goal of the "robotic pet—playing around with the robot, using hand gestures" behavior was to show that robots can be used as pets

that recognize different hand gestures. With the "Follow the leader" behavior, a group of robots follows tele-operated robot that holds an AR marker. By using AR markers, ClearBot can follow specific commands. The goal was to show that robots can work in groups (by tracking and following the lead robot with an AR marker), achieving more than only one robot (15 min).

- Participants tried these behaviors out by themselves and discussed with researchers on the applicability of these types of behaviors in a real-life context (30 min).
- ClearBot robotics platform was presented in depth by its developers. The need for the ClearBot robot and its main mission (close the gap between the simple educational robots and advanced industrial robots) was explained to participants. During the presentation, the developers gave an extensive overview of the robot's hardware components and technical capabilities, which were exemplified by a selection of demo behaviors (15 min).

13.4 Results

13.4.1 Do Participants' Intentions About Collaboration with Robots in the Future Change After the Experience-Based Workshop?

We had asked the participants to describe their vision about robots at their homes and workplaces in ten years. In particular, they were asked to indicate whether they saw robots as their friends or foes, colleagues or family members, or simple tools.

The participants' **answers prior to the workshop** (Table 13.1) indicated that in workplace settings, the majority of participants (63% in total, or 67% of men and 57% of women) perceived robots as possible colleagues. 44% of participants (44% of men, 43% of women) viewed robots as tools; and 31% of participants (56% of men vs. 0% of women) considered robots to be friends. However, at home, the majority of participants (88% in total, or 89% of men and 88% of women) saw robots as tools only. A quarter of respondents (22% of men and 29% of women) considered robots as potential friends, and 25% of respondents (44% of men and 0% of women) perceived robots as potential house servants. Only one participant (a man) considered robots as family members.

The **answers after the workshop** (Table 13.1) reflected some changes in the attitudes of the participants. In the workplace settings, the robots were now considered mostly as tools (62%, or 67% of men and 57% of women) and somewhat as colleagues (54%, or 50% of men and 57% of women). Robots were also seen friendlier than before (38%, or 33% of men and 44% of women). At home, the robots were seen foremost as tools (92%, or 100% of men and 86% of women) or servants (38%, or 33% of men and 43% of women). Less than a quarter of participants would consider

Table 13.1 Participants' perception about the role of the robots in their lives after 10 years

		Workplace			Home			
		Tool (%)	Colleague (%)	Friend (%)	Tool (%)	Family member (%)	Friend (%)	Servant (%)
Pre	Men	44	67	56	89	11	22	44
	Women	43	57	0	86	0	29	0
	All	**44**	**63**	**31**	**88**	**6**	**25**	**25**
Post	Men	67	50	33	100	0	17	33
	Women	57	57	43	86	0	29	43
	All	**62**	**54**	**38**	**92**	**0**	**23**	**38**
Δ	Men	23	−17	−23	11	−11	−5	−11
	Women	14	0	43	0	0	0	43
	All	**18**	**−9**	**7**	**4**	**−6**	**−2**	**13**

Source: Bold values indicate the results of both genders combined

robots as friends at home (23%, or 17% of men and 29% of women), and none of the participants would view robots as family members.

13.4.2 What Are the Participants' Suggestions on Improvement of the Experience Workshop Design?

We asked participants to reflect on their experience of the workshop and suggest improvements to workshop design. Without exception, all respondents implied that the workshops had had a positive impact. In addition, some areas for improvement were identified. Many participants had no previous first-hand experience with robots or programming and found it difficult to construct a comprehensive understanding about the collaborative robots. The areas for improvement were indicated as follows: First, the materials should be designed to be used by the people with no previous experience with robotics—avoiding advanced terminology and coding, while providing more details in areas where beginners have less knowledge. Second, teamwork should be carried out in small groups, in order to provide participants with more individual and personal hands-on experience. Third, instead of starting with a theoretical overview, it is better to have a short concise introduction first. Analysis indicated that it could be also reasonable to conduct the hands-on experience part in the beginning of the workshop (before giving theoretical overview), as this could help participants to lay foundations for constructing their knowledge about robotics, for this is an area previously unknown to them. Fourth, participants suggested that the workshop materials should be based on the actual workshop content (we had used different robot behaviors in advertising materials and during the workshop).

Fifth, some participants pointed out that the workshop should also discuss potential malfunctions of the robots, as this will help to reduce the anxiety of interacting with robots.

13.5 Conclusions and Discussion

AI and ML are making robots more intelligent than ever before, allowing them to perform tasks that were previously exclusive to humans. For many people, these developments fuel the feeling uncertainty about their future. Studies have shown that people's attitudes depend on their previous experience with robots. This paper examined the effects of a two-day educational robotics workshop on people's attitudes about the role of robots in their workplaces and homes in imaginative future of 10 years from now, and the participants' recommendations for improving such workshops.

First, we found that our study participants (university students, average age 33 years) did not have any hostile feelings toward robots in the beginning. People were more receptive to the idea of having robots as colleagues (or even friends) in their workplaces, whereas in their private life (at home) they saw robots mostly as tools or servants. It seems that adding intelligence does not make robots equal partners to humans. Robot is mostly seen as a tool, perhaps a sophisticated tool, which has to be treated carefully—but nevertheless, it is a tool that is limited to its functions. In a way, this attitude is similar to how slaves were treated in ancient societies, or how feudal lords treated their serfs up to the nineteenth century.

Second, we saw remarkable differences between the attitudes of men and women. For example, before the workshop none of the female participants perceived robots as friends at their workplaces, or (semi-intellectual) servants at their homes—whereas after the workshop almost half of them recognized robots as potential work—friends or domestic servants. These results imply that after the workshops the female participants had a greater tendency to view robots as animated objects. At the same time, men were more optimistic about the human-like features of the robots before the workshop and less after it. Based on our findings, it seems that such workshops or similar activities should take into account gender differences. Perhaps other social and cultural differences need to be taken into account as well, but future research has to be carried out in order to deal with this question.

Third, it seems that workshops like the one described in this paper are successful in providing a better understanding about the limitations and opportunities of frontier technologies. On the one hand, such workshops are helpful in preparing us for the life that involves technological alterities. On the other hand, they facilitate us to make up our minds about the role of robots in our workplaces and homes. We regard this an important aspect of human resolve that might help to oppose the possible pressure of corporations either to replace people in the workplace altogether, or to force them to work as if they were themselves just robots [35]. Other similar studies have also shown that training increases people's technical and pedagogical knowledge about the development and adaptation of robots [36].

In addition, based on the participants' opinions on potential improvements of future workshops or similar activities, we argue that their content should be specifically designed for people with little or no previous experience with advanced robots. This means that all the participants should be initially provided with basic knowledge and some hands-on experience that helps them to form their understanding about ML, AI, and robots. As the presenters themselves have usually advanced knowledge about these topics, it would be good to include some inexperienced people to the workshop development process to see if the concepts and the ideas that are used for the explanations can be easily comprehended. For example, one of the people who has just carried out the workshop can participate in the organization of the next one, providing first-hand information about the technical deficiencies of the participants.

This study has some limitations that need to be addressed for initiating further research. First, for more accurate and generalizable results we recommend using a larger and more diverse sample, and conducting more iterations with robots. Second, we suspect that the type of robot could greatly determine the outcome of the study. We used Poppy Ergo Jr that is an educational model of robotic hand and often used in industry, and ClearBot that is somewhat similar to warehouse robots, such as used by the Amazon, for example. However, using humanoid social robots that are able to uphold simple conversations could lead to different results [37].

Acknowledgements Project "TU TEE—Tallinn University as a promoter of intelligent lifestyle" (No. 2014-2020.4.01.16-0033) under activity A5 in the Tallinn University Centre of Excellence in Educational Innovation.

Ethical Approval All procedures performed in studies involving human participants were in accordance with the ethical standards. Informed consent was obtained from all participants included in the study.

References

1. Hawking, S., Papathanasiou, V.: "It Can Be Done"—An Earth Day Message. The European Space Agency (2018)
2. United Nations: World Economic and Social Survey 2018. New York (2018)
3. DiMaio, S., Hanuschik, M., Kreaden, U.: The da Vinci surgical system. In: Rosen, J., Hannaford, B., Satava, R. (eds.) Surgical Robotics. Springer, Boston, MA (2011). https://doi.org/10.1007/978-1-4419-1126-1_9
4. Bugmann, G., Siegel, M., Burcin, R.: A role for robotics in sustainable development?, pp. 1–4. https://doi.org/10.1109/AFRCON.2011.6072154 (2011)
5. Manyika, J., Chui, M., Miremadi, M., Bughin, J., George, K., Willmott, P., et al.: Harnessing Automation for a Future That Works. McKinsey & Company (2017)
6. Gnambs, T., Appel, M.: Are robots becoming unpopular? Changes in attitudes towards autonomous robotic systems in Europe. Comput. Hum. Behav. **93**, 53–61 (2019)
7. Stein, J.P., Liebold, B., Ohler, P.: Stay back, clever thing! Linking situational control and human uniqueness concerns to the aversion against autonomous technology. Comput. Hum. Behav. **95**, 73–82 (2019)

8. Dang, J., Liu, L.: Robots are friends as well as foes: ambivalent attitudes toward mindful and mindless AI robots in the United States and China. Comput. Hum. Behav. **115**, 106612 (2021)
9. Jha, S., Topol, E.J.: Adapting to artificial intelligence: radiologists and pathologists as information specialists? JAMA **316**(22), 2353–2354 (2016)
10. Sauppé, A., Mutlu, B.: The social impact of a robot co-worker in industrial settings. In: Proceedings of the 33rd Annual ACM Conference on Human Factors in Computing Systems—CHI'15, pp. 3613–3622. https://doi.org/10.1145/2702123.2702181 (2015)
11. Reich-Stiebert, N., Eyssel, F., Hohnemann, C.: Involve the user! Changing attitudes toward robots by user participation in a robot prototyping process. Comput. Hum. Behav. **91**, 290–296 (2019)
12. Fraune, M.R., Kawakami, S., Sabanovic, S., De Silva, P.R.S., Okada, M.: Three's company, or a crowd?: the effects of robot number and behavior on HRI in Japan and the USA. In: Robotics: Science and Systems (2015)
13. Wurhofer, D., Meneweger, T., Fuchsberger, V., Tscheligi, M.: Deploying robots in a production environment: a study on temporal transitions of workers' experiences. INTERACT (2015). https://doi.org/10.1007/978-3-319-22698-9_14
14. EU Robotics AISBI: Robotics 2020—Strategic Research Agenda for Robotics in Europe. Years 2014–2020. SPARC. The Partnerships for Robotics in Robotics (2015)
15. Leite, I., Martinho, C., Paiva, A.: Social robots for long-term interaction: a survey. Int. J. Soc. Robot. **5**(2), 291–308 (2013)
16. Turja, T., Van Aerschot, L., Särkikoski, T.: Finnish healthcare professionals' attitudes towards robots: reflections on a population sample. Nurs. Open **5**(3), 300–309 (2018)
17. Eagly, A.H., Chaiken, S.C.: Attitude structure and function. In: Gilbert, D.T., Fiske, S.T., Lindzey, G. (eds.) The Handbook of Social Psychology, pp. 269–322. McGraw-Hill, New York (1998)
18. Fazio, R.H.: Multiple processes by which attitudes guide behavior: the ODE model as an integrative framework. In: Zanna, M.P. (ed.) Advances in Experimental Social Psychology, pp. 75–109. Academic Press, San Diego (1990)
19. Fazio, R.H., Roskos-Ewoldsen, D.R.: Acting as we feel: when and how attitudes guide behavior. In: Brock, T.C., Green, M.C. (eds.) Persuasion: Psychological Insights and Perspectives, pp. 41–62. Sage, Thousand Oaks (2005)
20. Brooks-Harris, J.E., Stock-Ward, S.R.: Workshops: Designing and Facilitating Experiential Learning. Sage, USA (1999)
21. OECD: The Future of Education and Skills. Education 2030. OECD, Paris (2018)
22. Belay, H., Ó Ruairc, B., Guérandel, A.: Workshops: an important element in medical education. BJPsych Adv. **25**(1), 7–13 (2019)
23. Knowles, M.S., Holton, E.F., Swanson, R.A.: The Adult Learner: The Definitive Classic in Adult Education and Human Resource Development, 6th edn. Butterworth-Heinemann (2012)
24. Leoste, J., Heidmets, M., Ley, T.: What makes new technology sustainable in the classroom: two innovation models considered. In: Mealha, Ó., Rehm, M., Rebedea, T. (eds.) Ludic, Co-design and Tools Supporting Smart Learning Ecosystems and Smart Education. Smart Innovation, Systems and Technologies, vol. 197. Springer, Singapore. https://doi.org/10.1007/978-981-15-7383-5 (2021)
25. Clarke, R.: Principles and business processes for responsible AI. Comput. Law Secur. Rev. **35**(4), 410–422 (2019)
26. Wisskirchen, G., Biacabe, B.T., Bormann, U., Muntz, A., Niehaus, G., Soler, G.J., von Brauchitsch, B.: Artificial Intelligence and Robotics and Their Impact on the Workplace. IBA Global Employment Institute (2017)
27. Mitchell, T.: Machine Learning. McGraw Hill, New York (1997)
28. Peshkin, M., Colgate, J.E., Wannasuphoprasit, W., Moore, C., Gillespie, B., Akella, P.: Cobot architecture. IEEE Trans. Robot. Autom. **17**(4), 377–390 (2001)
29. Kawamoto, H., Sankai, Y.: Power assist system HAL-3 for gait disorder person. In: Miesenberger, K., Klaus, J., Zagler, W. (eds.) Computers Helping People with Special Needs, pp. 196–203. Springer, Berlin Heidelberg (2002)

30. Rex Bionics—Company History. http://www.rexbionics.com/about-us/company-history/. Accessed 14 Mar 2021
31. El Zaatari, S., Marei, M., Li, W., Usman, Z.: Cobot programming for collaborative industrial tasks: an overview. Robot. Auton. Syst. **116**, 162–180 (2019). ISSN 0921-8890
32. Huxley, K.: Content analysis, quantitative. In: Atkinson, P., Delamont, S., Cernat, A., Sakshaug, J.W., Williams, R.A. (eds.) SAGE Research Methods Foundations (2020)
33. Smith, J.A., Flowers, P., Larkin, M.: Interpretative Phenomenological Analysis: Theory, Method and Research. Sage, London (2009)
34. National Academy of Engineering: Sensing and Shaping Emerging Conflicts: Report of a Workshop by the National Academy of Engineering and United States Institute of Peace Roundtable on Technology, Science, and Peacebuilding. The National Academies Press, Washington. https://doi.org/10.17226/18349 (2013)
35. Gutelius, B., Theodore, N.: The Future of Warehouse Work: Technological Change in the U.S. Logistics Industry. UC Berkeley Labor Center; Working Partnerships USA (2019)
36. Coşkun, T.K.: The effectiveness of robot training in special education: a robot training model proposal for special education. Interact. Learn. (2020)
37. Silva, K., Lima, M., Santos-Magalhães, A., Fafiães, C., de Sousa, L.: Living and robotic dogs as elicitors of social communication behavior and regulated emotional responding in individuals with autism and severe language delay: a preliminary comparative study. Anthrozoös **32**(1), 23–33 (2019). https://doi.org/10.1080/08927936.2019.1550278

Part IV
Methods, Processes and Communities

Chapter 14
Co-design and Shared Practices: An Overview of Processes of Learning in Nonformal Educational Contexts

Inês Santos Moura⬭ and Vania Baldi⬭

Abstract The present article is an overview that outlines co-design methodology in constructing an informal process of contextually situated and encouraged knowledge. The shared practices are related to emancipation and participation processes that integrate co-design implementation, as is the case of photovoice. On the other hand, the co-design methodology includes strategies and tools that foster digital literacy in a nonformal education context. In this regard, it is intended to reflect how co-design can also produce knowledge about that social context. Participatory methodologies, as co-design, digital literacy, and collaborative learning, are essential tools for promoting digital inclusion. Therefore, "we may say that the anthropological transformations and mediamorphoses are interwoven, reflecting it in the practices of learning, knowledge and in participation in collective life" (Baldi in La comunicación especializada del siglo. McGraw-Hill/Interamericana, p. 591, 2020 [2]). In this context, it is essential to promote an inclusive smart education that will allow individuals to produce, access, and effectively communicate through technological means.

Keywords Co-design · Participation · Digital literacy · Nonformal education

14.1 Introduction

Designers have been getting closer to the users who enjoy their products over the last six decades, using various tools and strategies, as is the methodological approach of co-design that started in Northern Europe [20]. This design methodology emerged between the 1970s and 1980s in Scandinavia, intending to provide research work accomplished in partnership with workers' unions and academics. So that workers could determine the form and which new technologies would be introduced into the

I. S. Moura (✉) · V. Baldi
University of Aveiro, Aveiro, Portugal
e-mail: inessantosmoura@ua.pt

V. Baldi
e-mail: vbaldi@ua.pt

© The Author(s), under exclusive license to Springer Nature Singapore Pte Ltd. 2022
Ó. Mealha et al. (eds.), *Ludic, Co-design and Tools Supporting Smart Learning Ecosystems and Smart Education*, Smart Innovation, Systems and Technologies 249,
https://doi.org/10.1007/978-981-16-3930-2_14

environment. This approach to work and research aimed to empower the worker and foster democracy and participation in the workplace [21].

As the workers lacked knowledge and experience in ecosystem design (this includes people, objects, processes, interfaces, others that collaborate to propose solutions), the UTOPIA project emerged. With this project, it was possible to bring together researchers and union workers to experience and use various organizational tools, workshops, and other research techniques. The UTOPIA project was not satisfactory in creating a working system but succeeded in developing a design approach and various methods for a participatory design working methodology [21].

In participatory design or co-design, the roles are mixed, and the people who will use the services or products being developed also play an essential role in the development of knowledge, generation of ideas, development of the concept, and the product itself [20]. Consequently, the co-design methodology has participation as a significant element, and it is also relevant in the processes of digital learning in the context of nonformal education. These learning and participation processes are important to produce knowledge about a certain reality that is closely familiar and intended to be better known by suggesting intervention practices within it (Fig. 14.1).

Thus, digital literacy also plays an important role in learning and cognitive processes, as Conrad explains "Digital literacy is changing the way we think, for better or worse" [8, p. 375]. Some studies indicate that the use of digital media "as reading tools may diminish many of the cognitive processes necessary for reading comprehension and learning in general" [8, p. 371]. But simultaneously, digital literacy can promote a new mindset improving the cognitive abilities to succeed "in today's knowledge-based society" [8, p. 371].

Fig. 14.1 Various contexts and elements of the co-design methodology facilitate the participatory and production of knowledge processes

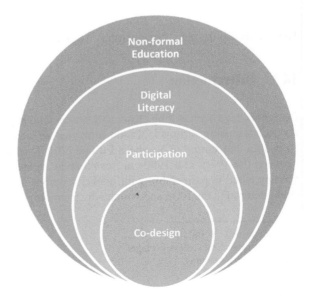

14.2 Co-design and Digital Literacy

In the contemporary social and technological context, digital literacy is relevant because it allows the individual to read, interact, and critically understand the info-communication contents and the respective creation and production mechanisms to which he/she has access. As Castells [7] comments, new information technologies are not just tools to be used but also developed processes. Users of technology can simultaneously be creators, and therefore, "users can take control of technology as in the case of the Internet" [7, p. 69]. Individuals produce, use, and access information and communicate through new technological means, producing new ways of sharing and communicating in networks. As researchers, it will be essential to know how to lead and facilitate the creative expression of the people involved, and users are part of the design team as experts of their experiences, and for this, they must have access to the appropriate tools so that they can participate [20].

Castells [7] comments that there is a growing relationship between mind and machines, and that, because of this, there is an alteration in the way we live our daily lives, how we learn, work, produce, and dream. Castells considers that "computers, communication systems, decoding, and genetic programming are all amplifiers and extensions of the human mind" [7, p. 69]. As Conrad comments, "digital literacy refers to a shift in the way we think" [8, p. 373]. Nowadays, it is possible to acknowledge that information and communication technology (ICT) has a great effect and influence on our lives [12]. From this regard, "ICT has changed the way in which people are informed, communicate, and collaborate" [12, p. 1].

Considering the changes that are present in a "socio-technical scenario of digital connectivity, there come psycho-social changes linked to a different dimension and conception of temporality, intimacy, curiosity, environmental sensitivity itself, and self-perceptive" [2, p. 591].

The relationship of the individual with society, computer, communication, and information systems enables a reflection and analysis of digital literacy's importance to create and apply these dynamics. The creation and sharing of information and knowledge will be more effective and productive if there are, in an info-communication context of design and use, individuals who can think, understand, and use in greater depth the info-communication tools at their disposal.

Digital literacy is also an essential tool for promoting digital inclusion. Individuals who use and understand without difficulty the different aspects related to info-communication processes in digital and technological environments "put themselves at an advantage not only in terms of education and employment—the items most highlighted in public policies—but in all aspects of life that require communication and information" [5, p. 292].

Castells [7] states that many areas of the world and population segments are isolated and disconnected from this new technological system, and "the disconnected areas are culturally and spatially discontinuous: They are in the inner cities of the US or the suburbs of France, as well as in the African slums and the Chinese and Indian rural deprived areas" [7, p. 70]. Castells [7] also explains that through this phenomenon of disconnection, social and technological inequality proceeds. As Norris explains, "technological opportunities are often highly unevenly distributed, even in nations like Australia, the US, and Sweden at the forefront of the information society" [18, p. 1].

It is possible to observe that the movement of technological spread is not comprehensive and inclusive. Digital literacy may be a relevant element to counteract digital exclusion and provide individuals with a critical understanding of information and communication mechanisms, promoting contact with the info-communication technological instruments and environments/contexts. Digital literacy enables constructive social action, critical/creative thinking, and innovation in the field of information and communication technology. The building blocks of digital literacy are more than computer skills. They encompass more complex capabilities such as understanding, designing, analyzing, and creating different content using various digital tools. Therefore, it is critical to promote skills that are not only about *technical abilities*. Digital literacy enables individuals to participate, contribute, and benefit more fully in the global and digital society. Individuals become not only consumers but also active producers and contributors to a digital community. The creation of the digital environment enables the development of skills in the fields of, for example, citizenship, technological appropriation, cultural empowerment, and information search. In this way, it also enables social awareness, sharing of knowledge, and evaluation skills. The literacy process is demarcated and influenced by each community's social, political, and cultural contexts in which it is inserted [5]. Therefore, this learning process should be thought and applied concerning its multidimensional quality.

Relating to co-design experiences, Sanders comments that "the roles of the designer and the researcher blur and the user becomes a critical component of the process. The new rules call for new tools. People want to express themselves and to participate directly and proactively in the design development process" [19, p. 2]. Therefore, there is a collaborative working process in developing ideas and concepts, creating and evaluating sketches, and reflecting on the models and prototypes [22]. Figure 14.2 shows the different practices used in developing products and services, taking human-centered design into account.

The co-design methodology provides users and designers with access to different materials and techniques to express their memories, feelings, reflections, and visions envision a future situation [22]. In this sense, co-design has developed methods to include and encourage users' participation and other critical elements in the design processes [1]. The tools used also help facilitate communication within the project team and support the construction and strengthening of a relationship of inspiration, empathy, and involvement with the participating users [22]. From this point of view, participation is an element present in the co-design methodology, and the objective is that the users of the technology can make decisions collectively about a given product and context. Furthermore, "participating demands specific actions of skilling and tooling that citizens need to undertake to equip themselves" [13, p. 81]. The authors Jenkins et al. [14] reveal that participation is not only focused on DIY production (*"Do it Yourself"*) but also on other types of actions—evaluation, criticism, and recirculation of material. As they state:

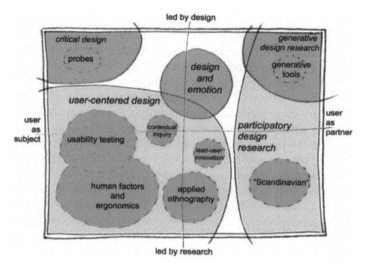

Fig. 14.2 Different practices of design. *Source* Article "Co-creation and the new landscapes of design" of Sanders and Stappers [20]

> We believe that there are still people who are primarily just "listening to" and "watching" media produced by others. However, like Yochai Benkler (2006), we argue that even those who are "just" reading, listening, or watching do so differently in a world where they recognize their potential to contribute to broader conversations about that content than in a world where they are locked out of meaningful participation. [14, pp. 154–155]

According to Jenkins et al. [14], there are two conceptions of participation. The first concept is related to a corporate perspective—companies aim to respond more effectively to their consumers' needs and desires. The other conception of participation refers to a political view. The primary objective is the possibility of everyone, as citizens, to exercise power over decisions that impact and transform their quality of life. Some activist groups seek to transform audiences into public networks to promote their causes [14].

Jenkins et al. [14] explain that "today, academics are much more likely to talk about politics based on "participation," reflecting a world where more media power rests in the hands of citizens and audience members, even if the mass media holds a privileged voice in the flow of information" [14, p. 163].

Different communities have worked together to broaden participation and overcome restrictions in their communication capacities. Some examples of this are the immigrant rights struggles in Los Angeles or the "Tecnobrega" movement in Brazil, "all suggest ways that our public sphere has been enriched through the diversification of who has the means to create and share culture" [14, p. 193].

Authors Jenkins et al. [14] explain that our culture is becoming more participatory concerning older mass communication systems. The ability for expressive and meaningful participation in online public networks is related to everyone's educational and economic opportunities, and thus, "the struggle over the right to participation is linked to core issues of social justice and equality" [14, p. 194].

Co-design distances itself from other approaches insofar as the design work must be elaborated not only considering its users' needs. Still, it should be developed together with its intervening party (users, designers, researchers, among others), thus, creating a more comprehensive participatory work dynamic [21]. Participation and democracy are the central values of co-design, in this way, favoring a design practice based on a balanced power relationship, also giving users a "voice" in the development of activities and decision-making [22] (Table 14.1).

Co-design has become a relevant approach for researchers interested in human–computer interaction and computer-supported cooperative work [21]. The participatory design methodology is also used in urban planning, architecture, disadvantaged communities, and sustainable development [23]. Thus, it has also contributed to some social changes such as "the development of new technologies since the early days of participatory design, such as transformations in the socio-economic makeup of societies, development of personal technologies, the diffusion of information technology to every aspect of everyday life; and most importantly, more knowledgeable co-designers" [21, p. 165] (Fig. 14.3).

Table 14.1 Traditional design practices and emerging design practices

The traditional design disciplines focus on the **designing of** "products" …	… while the emerging design disciplines focus on **designing for** a purpose
Visual communication design	Design for experiencing
Interior space design	Design for emotion
Product design	Design for interacting
Information design	Design for sustainability
Architecture	Design for serving
Planning	Design for transforming

Source Article "Co-Creation and the new landscapes of design" by Sanders and Stappers [20]

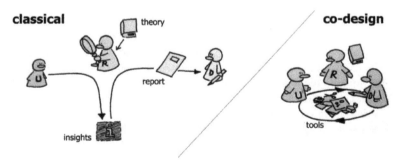

Fig. 14.3 On the left, the traditional roles of users, researchers, and designers in the design process, and on the right, the parts merge in the co-design process. *Source* Article "Co-Creation and the new landscapes of design" by Sanders and Stappers [20]

It can be stated that there are three stages present in the investigation of co-design; the first stage concerns the initial exploration of the work. In this period, the intervening group, together with the researchers, begins a familiarity process, intending to survey the technologies they use. Various working methods are used in this stage's development, such as participant observations and interviews [21]. Participatory audiovisual methodologies may also be used in this initial phase of work development, as photovoice. The participatory methodology of photovoice was first presented by Wang and Burris as a Photo novella in 1994 and has since become a methodological empowerment tool that allows participating individuals to reflect on their communities' potentials and concerns.

Photovoice takes place in three stages:

> Individuals go into their communities and take pictures of their concerns. Once completed, the individuals move onto facilitated discussions, sharing with one another what the photographs mean to them. The group dialog allows the individuals to build upon each other's concerns, helping shape the identified needs of the community. [15, p. 2]

Photography helps to legitimize acts, situations, and contexts, allowing the construction of subjects' representation and stereotypes' rupture [4]. Researchers also recognize photovoice as an essential tool for community-based participatory research for obtaining accurate and correct information collection [11, 15].

The next step is entitled the discovery process, i.e., this phase is related to the clarification and identification of the goals that users intend to achieve at the end of the project. The working process between users and researchers relies on some methodologies such as, for example, the construction of a storyboard and the creation of workflow organization models [21].

The third stage deals with prototypes' construction, promoting the project ideals' materialization in collaboration with users and researchers. Some techniques to be used are constructing mock-ups and paper prototypes in cooperation, among others [21].

14.2.1 Processes of Learning: Making Things Together and Nonformal Education

Learning is also doing things together [16], and the process of learning "can take place anytime and anywhere via the utilization of smart devices" [24, p. 4]. Gauntlett [9] explains that when we create something, we establish a form of connection, engaging in a social dimension and connecting with other people. Equally, when we make and share something in the world, we increase our engagement and connection with our social and physical environments. Gauntlett [9] indicates that by making something, we are engaged in the process of discovery and idea creation, "taking time to make something, using hands, gave people the opportunity to clarify thoughts or feelings, and to see the subject-matter in a new light" [9, p. 4]. Therefore, "making is about understanding in practice, and it is inextricably tangled with an active engagement with the material world and culture" [3, p. 25]. Gauntlett's [9] reflections are essential for reflecting on collaborative working processes. They demonstrate the importance of creating content, from images to other physical objects, to stimulate communication and direct connection with other people by presenting and discussing them.

From this perspective, nonformal education helps a particular community develop their skills to improve their lives and their relationships in society [3]. Therefore, it can be used as a form of learning within the community to identify what it would like to improve or change using its resources and learning new skills (digital or other) to achieve this [3].

Following on from this, "every activity or product developed from nonformal education can be tangible, but the sustainability of the activity or the product itself lies in the skills and abilities the community has gained through the process" [3, p. 9].

In cities, there is a strong relationship between the role of citizens and technology, which enables transformations in their form and how they operate [6]. Cities "are centers of social and political life where not only wealth but also knowledge, techniques, and works (works of art, monuments) accumulate" [17, p. 12].

Author Briones [6] explains that:

> The physical structure of cities (the hardware) is increasingly been intervened by new actors or actors with new roles. In the last years, new organizations of citizens are taking a leading role in the urban scenario, working collaboratively among them and with local institutions for achieving solutions to their social needs. [6, p. 3263]

In this regard, "smart education under smart city architecture is to provide every citizen personalized services and seamless learning experience" [24, p. 15]. Also, nonformal education contains a participatory learning process within it. It is a "grass-roots approach helping people to clarify and address their own needs, taking an active role to fulfill them" [3, p. 9]. Thus, the learning process's focus is the learners' needs [3]. Nonformal education assumes a relevant role in the constitution and the creation of a citizen practice exercised by people from any social stratum [10].

This leads back to Fig. 14.1 which illustrates that in learning processes, participation is also present. This participation can be seen as a form of citizen and civic participation, where people get involved in making something, whether it is a mobile app, a movie, or another object. These learning processes also involve working toward a common purpose.

Figure 14.4 shows how several elements compose the learning processes and the construction of knowledge about and from social reality.

Nonformal education aims at the formation of emancipated citizens, thus also promoting citizen education [10]. So, when people participate, they tend to continue in this process of civic involvement and their active participation grows [10].

Fig. 14.4 Some of the elements of participatory learning processes and nonformal education

14.3 Conclusions

With the revolution in information technology and the technological paradigm addressed by Castells [7], individuals have become participating subjects who search, investigate, analyze, and reflect on how to use the information available in various technological means of communication, such as the Internet. In this order, the "digital revolution represents the fluid infrastructure of our lives" [2, p. 591]. For this reason, the individual will have to be equipped with tools that make him capable of analyzing, understanding, communicating, and using the information that is present in his daily life. So, it is necessary to establish a conversational connection to reflect and seek to understand the networks and intersections that exist and could be established in a nonformal educational environment that promotes digital literacy, participation, considering the current technological and social phenomena.

The co-design methodology is a way to create "glances" at the social context, considering all in the creative process of sharing and generating knowledge and creating solutions. The promotion of digital inclusion and info-communication literacy aids a technological evolution that could influence and transform society.

Acknowledgements Research project funded by the Fundação para a Ciência e a Tecnologia (FCT), with the Individual Doctoral Scholarship—SFRH/BD/131706/2017.

References

1. Andersen, L.B., Danholt, P., Halskov, K., Hansen, N.B., Lauritsen, P.: Participation as a matter of concern in participatory design. CoDesign **11**, 3–4 (2015)
2. Baldi, V.: Interpassive communication the age of automatized culture: new challenges for the ethics of mediation. In: Vicente Domínguez, A., Abuín Vences, N. (eds.) La comunicación especializada del siglo XXI, pp. 591–605. McGraw-Hill/Interamericana (2020)
3. Baldi, V., Filcakova, A., Gallucio, P., Matorčević (ex Jurić), D.: To Be or Not to Be... Formal. Tools for Recognition of Nonformal Education in Youth and Voluntary Work. Associazione InformaGiovani, Palermo, Italy (2016)
4. Banks, M.: Visual Methods in Social Research. Sage (2001)
5. Borges, J., Oliveira, L.: Competências infocomunicacionais em ambientes digitais. Observ. (OBS*) J. **5**(4), 291–326 (2011)
6. Briones, M.A.: Information design for supporting collaborative communities. Des. J. **20**(sup1), S3262–S3278 (2017)
7. Castells, M.: A Sociedade em Rede, vol. 1. Paz e Terra, São Paulo (1999)
8. Conrad, N.J.: Is digital literacy changing the way we think? L'Année Psychol. **118**(4), 371–376 (2018)
9. Gauntlett, D.: Making is Connecting: The Social Meaning of Creativity, from DIY and Knitting to YouTube and Web 2.0. Polity Press (2011)
10. Gohn, M.G.: Educação Não Formal, Aprendizagens e Saberes em Processos Participativos. Investigar em Educação—IIª Série n°1 (2014)
11. Graziano, K. J.: Oppression and resiliency in a post-apartheid South Africa: unheard voices of Black gay men and lesbians. Cult. Divers. Ethn. Minor. Psychol. **10**(3), 302–316 (2004)
12. Hennig, S., Vogler, R.: User-centred map applications through participatory design: experiences gained during the 'YouthMap 5020' project. Cartogr. J. 1–17 (2016)
13. Isin, E., Ruppert, E.: Being Digital Citizens. Rowman & Littlefield International, Ltd. (2015)
14. Jenkins, H., Ford, S., Green, J.: Spreadable Media: Creating Value and Meaning in a Networked Culture (2013)
15. Kuratani, D., Lai, E.: TEAM Lab—Photovoice Literature Review. Team Lab—Tobacco Education and Materials LAB (2011). http://teamlab.usc.edu/learn/literature-reviews.html
16. Lave, J., Wenger, E.: Situated Learning: Legitimate Peripheral Participation. Cambridge University Press, Cambridge (1991)
17. Lefebvre, H.: O Direito à cidade. Centauro Editora (2011)
18. Norris, P.: The digital divide. In: Webster, F., Blom, R., Karvonen, E., Melin, H., Nordenstreng, K., Puoskari, E.: The Information Society Reader. Routledge, London (2020)
19. Sanders, E.B.: From user-centered to participatory design approaches. In: Frascara, J. (ed.) Design and the Social Sciences Making Connections. Taylor & Francis Books Limited, London (2002)
20. Sanders, E.B., Stappers, P.J.: Co-creation and the new landscapes of design. Co-design **4**, 5–18 (2008)
21. Spinuzzi, C.: The methodology of participatory design. Tech. Commun. **52** (2005)
22. Steen, M., Kuijt-Evers, L., Klok, J.: Early user involvement in research and design projects— a review of methods and practices. In: 23rd EGOS Colloquium (European Group for Organizational Studies), pp. 5–7. Vienna (2007)
23. van der Velden, M., Mörtberg, C.: Participatory design and design for values. In: van den Hoven, J., Vermaas, P., van de Poel, I. (eds.) Handbook of Ethics, Values, and Technological Design. Springer, Dordrecht (2015)
24. Zhu, Z., Yu, M., Riezebos, P.: A research framework of smart education. Smart Learn. Environ.**3** (2016). https://doi.org/10.1186/s40561-016-0026-2

Chapter 15
Fostering Co-UXers in Later Age: Co-designing a UX Toolkit for the Senior Online Community MiOne

Sónia Machado⑩, **Liliana Vale Costa**⑩, **Óscar Mealha**⑩, **Ana Veloso**⑩, and **Carlos Santos**⑩

Abstract Over the past few years, an increasing dependence of the general population on information and communication technologies has been observed. As such, inclusive and age-friendly design has been of utmost importance to meet the context of every age cohort. Most of the software development projects tend to focus on the assumption relative to the users' cognitive models, and when addressing senior citizens, these are very likely to have the risk of falling into stereotypes and non-used products. By contrast, involving senior citizens in the early stages of the design process tends to be crucial. The aim of this paper is to identify and validate a co-design process capable of influencing senior citizens' experience in online communities. One of the elements of the co-design process that is described in the paper is a co-design toolkit to foster positive user experiences when interacting with online communities. The co-design process involved 36 adult learners from five Universities of the Third Age and 2 coordinators, organized into a set of 17 group sessions in the context of the miOne online community. Besides this co-design toolkit, a set of 'best practices' for involving the users in the co-design process and implications on the redesign of a Senior Online Community are proposed.

Keywords Co-design · User experience · Senior citizens · Online communities

S. Machado · L. V. Costa (✉) · Ó. Mealha · A. Veloso · C. Santos
Department of Communication and Art, University of Aveiro—DigiMedia, Aveiro, Portugal
e-mail: lilianavale@ua.pt

S. Machado
e-mail: scsm@ua.pt

Ó. Mealha
e-mail: oem@ua.pt

A. Veloso
e-mail: aiv@ua.pt

C. Santos
e-mail: carlossantos@ua.pt

Ó. Mealha et al. (eds.), *Ludic, Co-design and Tools Supporting Smart Learning Ecosystems and Smart Education*, Smart Innovation, Systems and Technologies 249,
https://doi.org/10.1007/978-981-16-3930-2_15

189

15.1 Introduction

Research on co-designing user's experiences has recently been receiving attention in such areas as marketing and tourism research [1]. Although many studies engage senior citizens in co-design activities (e.g., [2, 3]), few studies (e.g., [4]) use the approach of co-designing experiences (co-UXers) and engagement in online platforms. Indeed, the concept co-UXers is used in this paper to refer to the collaboration between the development team and the end users in the whole process.

Given this mentioned lack of research in the area of human–computer interaction (HCI), the aim of this study is to understand the way a co-design process can influence the experience of senior citizens in online communities, by engaging senior citizen participants of Universities of the Third Age with a toolkit of co-design activities to test and enhance their user experience within an online community context, i.e., a Portuguese online community entitled miOne, aimed at fostering active ageing.

It is worth also mentioning that Universities of the Third Age are organizations aiming to provide learning for older people, having classes and other activities for people aged 50 and over, being the target group of the miOne online community.

This study was developed in the context of the SEDUCE 2.0[1] POCI-01-0145-FEDER-031696 (SEDUCE. 2.0—Senior Citizen Use of Communication and Information in miOne community), in which the project's main goals are [5]: (a) to assess the impact of psychosocial variables and the sociability of senior citizens through the use of ICT in the context of the miOne online community[2] and (b) contribute to the growing development of the miOne online community with the participation of senior citizens.

15.2 Senior Citizens as Co-UXers of Design Processes

The concept of user experience (UX) has been adopted as a design goal, a philosophy, or/and method by many designers [6]. As a matter of fact, the increasing advances in information and communication technologies and its omnipresence and ubiquity in the end users' daily life have led to increased interest in UX design and study through the lens of graphic design, interaction design, product design, marketing, among others [7]. Although UX has been closely interrelated with co-design over the past decades [8], transferring this process to environment ecosystems that meet the end users' needs and context is lacking [9]. In this sense, co-designing user's experiences may also contribute to the knowledge of product design, making the consumers experts on the matter as long as the designer aims to provide a good user experience and, therefore, involve the end users in the design process [10]. The designer's role, then, changes from being a mere translator of the virtual user's cognitive model to

---

[1] Available at: http://www.seduce.pt/ (Date accessed: 20-05-2020).

[2] Available at: https://www.mione.altice.pt/ (Date accessed: 20-05-2020).

a facilitator of co-design processes that often rely on the development of toolkits in the end users' ideation and expression [11].

In co-design, the expertise and daily-life practices of the end users are considered as well as the role and meaning of technology to support them [3]. Whereas product design and assessment are usually based on the users' cognitive model and the designer's assumption of their behaviors, co-design involves different stakeholders such as researchers, designers, developers, potential customers, and users in the process from its beginning [11–13]. According to Reddy and colleagues [14], involving different stakeholders and interfering with the design process can bring relevancy to the UX Research [14, 15]; provide a sense of ownership and accountability of the stakeholders relative to product design and development [14, 15]; and strengthen social networks and community involvement through the contact and collaboration between organizations and all stakeholders [13, 16]. In addition, a co-design experience often helps to unravel the end users' unmet or unrecognized needs that cannot be anticipated with previous assumptions of their behaviors and cognitive models [17].

15.3 Method

A qualitative action research method is used in this study to answer the research question 'In what way can co-design influence the senior citizens' experience with an online community?', embodying four phases: (1) Co-design Research and Co-Design Toolkit Development; (2) Application of the Co-design Toolkit with adult learners from Universities of the Third Age; (3) Assessment of the Co-design Toolkit Effectiveness, involving co-design group workshops and questionnaires; and (4) (Re)Design of the Online Community miOne and proposal of best practices and guidelines for designing senior online communities.

15.3.1 Phase 1—Co-design Research and Toolkit Development

A set of co-design sessions were planned by contacting the Universities of the Third Age, defining the goals of each session, activities, materials, and the tools used for data collection. Those co-design sessions were carried out by applying a first version of the developed co-design toolkit, encouraging the interaction with the online community miOne, focused on the main themes (cf. Fig. 15.1)—i.e., communication, news, tourism, and health; also taking into consideration the three phases of the online community use—pre-experience, experience, and post-experience.

This toolkit consists of seven activities, one for the pre-experience ('More than Words'), five for the experience phase ('Secret Rules', 'Lucky vs Unlucky Tourist',

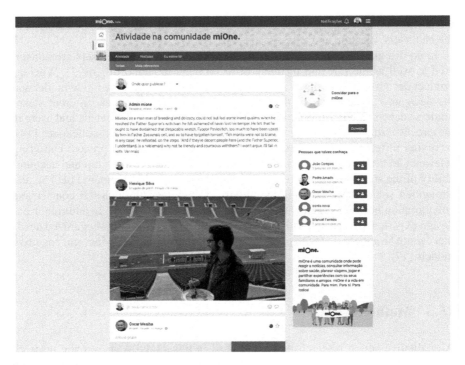

Fig. 15.1 miOne online community (main page)

'Secret Friend', 'Find the Fake News', 'News Match', 'Win-Win', and 'Doctor-Patient'), and one for the post-experience ('Pic-UX'). These activities were relative to the topics covered in the community (tourism, news, health) and enabled the identification of users' needs, functionalities, and interface (re)designing using, for example, concept maps, user journey maps, and AEIOU Registration (activities, environments, interactions, objects, and users) [18]. Furthermore, a sticker book was given at the beginning of the process, as part of the toolkit and as a gamification tool (c.f. Fig. 15.2) to ensure participants' weekly involvement in the activities.

15.3.2 Phase 2—Applying the Co-design Toolkit

Five Universities of the Third Age participated in the study in up to 6 co-design sessions per University of Third Age from February to May of 2020. The purpose of these sessions was to apply the co-design toolkit to perform a user interface and user experience evaluation of a Senior Online Community.

The criteria for this study sample were (a) being 50 years old and over; and (b) showing interest in ICT. The study sample consisted in up to 36 students per session,

Fig. 15.2 Toolkit materials: 'More than Words', 'Secret Rules', 'Win-Win', and sticker book

47.22% males ($n = 17$) and 52.78% female ($n = 19$), aged between 52 and 86 years old ($M = 69.06$ years; SD $= 7.40$). A convenience sample was used in this study, and therefore, data collection cannot be inferred to other contexts. Due to the COVID-19 outbreak, some sessions had to be adapted to the online context.

15.3.3 Phase 3—Assessing the Co-design Sessions and Toolkit

To support the data collected from the co-design sessions and toolkit application and gain further insights on participants' experience with it and the miOne community, an online questionnaire was distributed to the participants of the co-design sessions and a semi-structured interview in a group setting was conducted with the online participants.

15.3.4 Phase 4—(Re) Design of the Senior Online Community miOne

The last phase consisted in redesigning of the online community was conducted, considering the inputs of the co-design sessions, questionnaire, and online semi-structured interview. A set of 'best practices' or guidelines for designing senior online communities were formulated based on the same data collection methods, and the online community miOne was redesign accordingly.

15.4 Results of the Co-design Action Research

The following subsections report on the practices that were drawn from involving the co-UXers in the process and data was supported by participants' observations, group discussions, and in situ research team experience.

15.4.1 Practices for Involving Co-UXers in Co-designing a Senior Online Community

The co-design process presented in the Method section (Phase 1) enabled us to identify a set of practices for involving the users in the design of online communities. These were:

- Deliver the activities in the form of short-time workshop session followed by reflection periods with the interview and questionnaire surveying. The integration of scenario-building in these activities is important to enable participants' self-expression and brainstorming;
- Use autobiographical storytelling games. For example, the game 'Secret Friend' in which participants had to guess their secret friend in the online community acted as cultural probes to encourage senior citizens' self-expression of their own experiences; and
- Consider the people's past, present, and future lives in pre-, during, and post-design.

During the application of a Co-design Toolkit (Phase 2), the following recommendations were identified:

- Delineate a strategy for motivating the participants' attendance in the sessions, through for example, the use of a sticker book and session stickers;
- Plan group dynamics (e.g., collaborative and competitive challenges) and iteratively access the participants' expectations toward the activities and developed products;
- Involve the coordinators/professors at the Universities of the Third Age/caregivers, whenever possible to facilitate the interaction with the target group.

For assessing the co-design sessions and toolkit effectiveness (Phase 3), twenty-one participants were surveyed about the toolkit design activities. The majority emphasized they enjoyed the exchange of ideas, enthusiasm, and friendliness of other participants and the session moderators, as well as the possibility to learn ICT. Meeting new people and role-play activities were mentioned as an added value of the toolkit activities, as well as the involvement of different generations.

Furthermore, 61.9% of respondents stated that they would use miOne for their education. Nevertheless, in times of a COVID-19 pandemic, when the Universities of Third Age were closed, nine participants revealed they would continue to use the Senior Online Community miOne, whereas the remaining eight stated they would not continue. Among the reasons for not participating were task overload and laziness. Another participant said that he/she did not use miOne because he/she had difficulties using the platform. Participants also stated that the communication with other people and the knowledge gained with the activities was what motivated them. However, two participants noted factors that demotivated them. One participant mentioned the lack of adherence to the activities and the other felt that there was a lack of communication when these activities would be carried out. To improve the toolkit activities, participants suggested games with riddles, proverbs, and activities with music.

Finally, the (re) design of the Senior Online Community miOne followed the three-phase co-design process and its implications are described in the next section.

15.4.2 Implications on the (Re)design of the Senior Online Community miOne

The 'Best Practices' and 'Guidelines' for the general requirements of an online community were identified through the combination of the applied methods and grouped into three main categories: (a) information architecture design; (b) inter-action design; and (c) visual design. Having the following subsections: navigation, content, use of color, use of images, user feedback, and support. The mentioned 'Best practices' and 'Guidelines' are summarized in Table 15.1 [18].

Regarding the features and miOne themes implemented at miOne and tested at the co-design sessions: groups, chat, news, and health; a set of recommendations for designing and implementing these features were also created and are summarized in Table 15.2.

As for the redesign of the online community miOne, some aspects of the existent design were altered with the mentioned guidelines. During the co-design sessions the need to have more than one community emerged, so that coordinators and teachers at the universities of the Third Age could share the content of the classes only with their students. Figure 15.2 shows the design proposal resultant of this process.

A summary of the recommended features/functionalities is provided below:

- Registration/login: In the registry in a platform, show examples of the special characters needed to fulfill in the forms, e.g., 'The password should contain at least one special character (e.g. ? - ;)'; the possibility to see the password at the

Table 15.1 Best practices and guidelines for designing senior online communities

Information architecture design	Navigation	Show clearly where the user is Display the home button on every page High contrast and bolder navigation cues should be provided Navigation should be consistent Avoid double clicks Avoid drop-down and hover menus Use icons that correspond to senior citizens affordances Create redundancy—always display text in front of the icons; different ways to go to one place
	Content	Language should be simplified, clear, and consistent Only relevant information should be displayed Important information should be concentrated in the center of the screen
Interaction design	User feedback and support	Checking for spelling errors Clear and easy-to-follow feedback of errors should be provided A community tutorial should be provided, and users should be able to access it whenever they need it
Visual design	Use of color	Saturated colors should be used High contrast colors should be used
	Use of images and graphics	Avoid the use of meaningless images or graphics

Table 15.2 Recommendations for some of the features and miOne themes

Chat	Other forms of communication besides writing should be provided Possibility to delete the message after sending it Noticeable and simple notifications should be given when users receive new messages Feedback on the users' visualization of a message should be provided
News	Categorization by theme should be available Users should be able to read the news without leaving the community page Users should be guided through the process of sharing news for the first time The process of sharing news should be simplified
Health	Information about the health of the user should remain private Health-related issues should be approached carefully

login and register and suggestions on the format of the required password should
be provided.

- Profile: Add the name of the user to the top navigation, alongside the picture.
  Furthermore, add the menu 'More' ('Mais') to avoid the second click on the
  profile name/picture. Icons should also be suggestive of possible actions—e.g.,
  configuration of users' profiles, pencil used to the input of writing something
  about themselves.

- Navigation: Icons should be labeled, as shown in the redesigned proposal for the
  left menu. The publication input should be kept always open, with a pre-selected
  place to post. In the proposal users have the text 'Publish in' ('Publicado em') at the
  top of the drop-down menu where they can select the place they want to post. The
  drop-down menu is followed by an 'alter' ('alterar') link, to reinforce the action
  of changing the place to post and create redundancy. To facilitate readability, the
  area of the text has now higher contrast and aims to catch users' attention by using
  their names at the beginning of the pre-designed text, e.g., 'Hey Maria, write here
  a new post …'. The icons of adding pictures/video/files are now accompanied
  by the correspondent text to avoid mistakes. Lastly, like the publication area, the
  comments section has now higher contrast on the text input (Fig. 15.3).

In all, the practices for involving co-UXers in co-design of a Senior Online
Community and the implications on the (re)design of the Senior Online Community
miOne pave the way into the processes that communities and co-design in learning
can be approached.

15.5 Discussions

The conducted co-design sessions and toolkit application, perceived by direct observation and analyzing the questionnaire and semi-structured interview, allowed participants to talk to each other, meet new people while learning more about ICT and
helping to redesign the online community miOne. That was especially important
during the COVID-19 pandemic where most of them were isolated, as Universities
of The Third Age were closed. Furthermore, as the online community miOne is desti-
nated to be used in the context of education, the fact that most of participants mention
that they have continued using the platform even with the Universities closed can be
an indicator that the co-design sessions were efficient to increase their interest on the
platform.

The co-design process described in this paper differs from the mere prototype
evaluation in the sense that users self-express their motivations, behaviors, and interaction problems during the development process instead of solely measuring the
interaction problems and task performance or comparing A/B design versions. This
procedure is in line with the King's perspective [17] on the need to attend end users'
needs without blindly relying on the researchers' assumption of cognitive model and
behaviors.

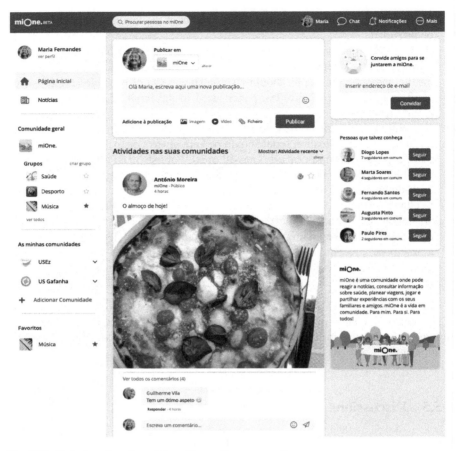

Fig. 15.3 Redesigned version of the miOne online community

The number of participants in this research may also bring some limitations on the guidelines for designing online communities, and hence, data may not be applied to other contexts.

Overall, the co-design process described in the paper is key to shape innovative learning ecosystems and community development, bringing a much more humanistic approach to the fore.

Acknowledgements This work was supported by the research project SEDUCE 2.0—Use of Communication and Information in the miOne online community by senior citizens. This project is funded by FCT—Fundação para a Ciência e a Tecnologia, I.P., COMPETE 2020, Portugal 2020 and European Union, under the European Regional Development Fund, POCI-01-0145-FEDER-031696 SEDUCE 2.0. Last but not least, a special thanks to all the participants that volunteered for this study and contributed with their time, expectations, and knowledge.

References

1. Mathis, E.F., Kim, H.L., Uysal, M., Sirgy, J.M., Prebensen, N.K.: The effect of co-creation experience on outcome variable. Ann. Tour. Res. **57**, 62–75 (2016). https://doi.org/10.1016/j. annals.2015.11.023
2. Ahmed, R., Toscos, T., Rohani Ghahari, R., Holden, R.J., Martin, E., Wagner, S., Daley, C., Coupe, A., Mirro, M.: Visualization of cardiac implantable electronic device data for older adults using participatory design. Appl. Clin. Inform. **10**, 707–718 (2019). https://doi.org/10. 1055/s-0039-1695794
3. Maa, S., Buchmuller, S.: The crucial role of cultural probes in participatory design for and with older adults. I-Com **17**, 119–135 (2018). https://doi.org/10.1515/icom-2018-0015
4. Wikberg-Nilsson, Å., Normark, J., Björklund, C., Axelsson, S.W.: HealthCloud: promoting healthy living through co-design of user experiences in a digital service. Proc. Nord. Des. Era Digit. Nord. **2018** (2018)
5. Veloso, A.I.: O que é o projeto SEDUCE? In: Veloso, A.I. (ed.) SEDUCE—utilização da comunicação e da informação em ecologias web pelo cidadão sénior. Edições Afronta-mento/CETAC.MEDIA, Portugal (2014)
6. Schifferstein, H.N.J., Hekkert, P.: Product Experience. https://doi.org/10.1016/B978-0-08-045 089-6.X5001-1 (2008)
7. Kou, Y., Gray, C.M.: A practice-led account of the conceptual evolution of UX knowledge. In: Conference on Human Factors in Computing Systems—Proceedings, pp. 1–13. https://doi. org/10.1145/3290605.3300279 (2019)
8. Wiklund, S., Cecilia, A., Åsa, B., Nilsson, W., Normark, J.: Participatory design of user interfaces for senior people's active aging (2018)
9. ASLERD: TIMISOARA DECLARATION: Better Learning for a Better World Through People Centred Smart Learning Ecosystems, pp. 1–9 (2016)
10. Kouprie, M., Visser, F.S.: A framework for empathy in design: stepping into and out of the user's life. J. Eng. Des. **20**, 437–448 (2009). https://doi.org/10.1080/09544820902875033
11. Sanders, E.B.-N., Stappers, P.J.: Co-creation and the new landscapes of design. CoDesign **4**, 5–18 (2008). https://doi.org/10.1080/15710880701875068
12. Pejner, M.N., De Morais, W.O., Lundström, J., Laurell, H., Skärsäter, I.: A smart home system for information sharing, health assessments, and medication self-management for older people: protocol for a mixed-methods study. J. Med. Internet Res. **21**, 1–9 (2019). https://doi.org/10. 2196/12447
13. Muller, M.J., Kuhn, S.: Participatory design. Commun. ACM **36**, 24–28 (1993). https://doi. org/10.1145/153571.255960
14. Reddy, A., Lester, C.A., Stone, J.A., Holden, R.J., Phelan, C.H., Chui, M.A.: Applying partic-ipatory design to a pharmacy system intervention. Res. Soc. Adm. Pharm. **15**, 1358–1367 (2019). https://doi.org/10.1016/j.sapharm.2018.11.012
15. Burns, K.E.A., Jacob, S.K., Aguirre, V., Gomes, J., Mehta, S., Rizvi, L.: Stakeholder engage-ment in trial design: survey of visitors to critically ill patients regarding preferences for outcomes and treatment options during weaning from mechanical ventilation. (2016). https:// doi.org/10.1513/AnnalsATS.201606-445OC
16. Jagosh, J., MacAulay, A.C., Pluye, P., Salsberg, J., Bush, P.L., Henderson, J., Sirett, E., Wong, G., Cargo, M., Herbert, C.P., Seifer, S.D., Green, L.W., Greenhalgh, T.: Uncovering the benefits of participatory research: implications of a realist review for health research and practice (2012). https://doi.org/10.1111/j.1468-0009.2012.00665.x
17. King, A.P.: Co-designing mobile collection points with older persons to promote green attitudes and practices in Hong Kong. Des. J. **22**, 1675–1686 (2019). https://doi.org/10.1080/14606925. 2019.1595000
18. Machado, S.: Co-designing a UX toolkit for senior online communities: the case of the online community miOne (2021). https://doi.org/10.13140/RG.2.2.19586.81602

Chapter 16
What Are the Latest Fake News in Romanian Politics? An Automated Analysis Based on BERT Language Models

Costin Busioc, Vlad Dumitru, Stefan Ruseti, Simina Terian-Dan, Mihai Dascalu, and Traian Rebedea

Abstract Social media and news outlets facilitate information sharing, while the Web is flooded by information posted online on a daily basis. However, content may be differently transmitted from case to case, based on the authors' intentions and vocabulary, to the extent that it generates completely opposite points of view. As such, fake news have become a global phenomenon, and recent events highlight a high impact of distorted or fake information, especially on the political side, when candidates' discourses include tendentious statements that require careful validation before completely trusting the source. This paper proposes an automated analysis of political statements in Romanian by applying different state-of-the-art Natural Language Processing techniques, and evaluating the importance of context in determining their veracity. Our corpus consists of entries from Factual, a recent Romanian fact-checking initiative that assembled a list of public statements, alongside relevant

C. Busioc · V. Dumitru · S. Ruseti · M. Dascalu · T. Rebedea
Department of Computer Science, University Politehnica of Bucharest, 313 Splaiul Independentei, 060042 Bucharest, Romania
e-mail: costin_eugen.busioc@stud.acs.upb.ro

V. Dumitru
e-mail: vlad.dumitru2212@stud.acs.upb.ro

S. Ruseti
e-mail: stefan.ruseti@upb.ro

T. Rebedea
e-mail: traian.rebedea@upb.ro

S. Terian-Dan
Department of Romance Studies, Lucian Blaga University of Sibiu, 10 Bld Victoriei, 550024 Sibiu, Romania
e-mail: simina.terian@ulbsibiu.ro

M. Dascalu (✉)
Academy of Romanian Scientists, 3 Str. Ilfov, 050044 Bucharest, Romania
e-mail: mihai.dascalu@upb.ro

Ó. Mealha et al. (eds.), *Ludic, Co-design and Tools Supporting Smart Learning Ecosystems and Smart Education*, Smart Innovation, Systems and Technologies 249,
https://doi.org/10.1007/978-981-16-3930-2_16

contextual information for their interpretation. Our results are comparable to similar experiments performed on the PolitiFact dataset, and represent a strong baseline for experiments in low-resource languages, like Romanian.

Keywords Natural Language Processing in Romanian · Fake news analysis · Transformer-based architecture

16.1 Introduction

Fake news include tendentious statements, pieces of public discourse that do not completely express facts or include exaggerated, baseless ideas introduced by speakers to strengthen a personal idea and for obtaining support, without being accurate. Fake news have become a global phenomenon [10], and a specific use case of high impact [1, 2] consists of political statements provided in different circumstances by either political candidates or politicians in office. Fact-checking in political discourses can be rigorously performed by providing clear explanations and interpretations. Two examples for such endeavours are PolitiFact[1] for US political statements and Factual[2] for Romanian political statements.

Concurrently with the spread of fake news that surpasses the sharing of facts [5, 20], there has also been a visible increase in the interest for research in Natural Language Processing (NLP) that would enable early detection of fake news and intervention to combat it. Moreover addition, spreading entirely fake news is not a common practice due to the ease with which the falseness can be detected and most information is partially altered. Consequently, managing to automatically understand the information shared in media and detecting potential segments that require attention, fact-checking and interpretation become essential for the process of news validation in political discourses and debates. One of the greatest challenges in automatically analysing and classifying fake information is obtaining a high-quality corpus of texts, with a rigorous annotation process performed by competent individuals in their fields who label specific sentences, larger fragments or entire articles. Furthermore, even if a specific collection is thoroughly analysed by experts, bias towards a specific interpretation may still exist due to the intrinsic subjectivity of certain texts. In addition, transferability of models between different datasets is limited as cross-domain analyses have lower accuracies than models trained on a given dataset [13].

Our goal is to introduce a baseline in terms of automated analyses of political statements in Romanian starting from Factual. Different state-of-the-art Natural Language Processing techniques are applied, while also accounting for the importance of context in determining the extent to which statements are valid or not.

The paper is structured as follows: the next section introduces various NLP architectures taken into consideration for fake news classification, followed by details on

---

[1] https://www.politifact.com/.

[2] https://www.factual.ro/.

our corpus, as well as a description of the method employed in our analysis that considers both the political statement and its context extracted from Factual. The fourth section presents our results, together with the confusion matrices for two models, while the fifth section provides in-depth discussions and describes limitations. Conclusions and future research initiatives are outlined in the last section.

16.2   Related Work

Assessing authenticity becomes a challenging task nowadays, especially when certain individuals tend to use the power of mass-media channels to manipulate the public, while pursuing political or financial gains. The negative effects of fake news are well known; people are exposed to the same erroneous statements and information from multiple sources to such an extent that they tend to give credence to and assimilate false information, making increasingly difficult to redress the problem afterwards [19]. Therefore, numerous studies analyse alternative ways of combating fake news throughout various online communities.

Wang [21] addresses the importance of confronting statements against their truth value using PolitiFact by developing a new fake news detection model that integrates text and meta-data. The experiments conducted in order to identify a false clause showed that a higher performance can be achieved expanding the search to include not only the statements themselves, but also part of the context and justification. The dataset introduced by the article is called the LIAR dataset, and it became one of the most popular datasets used in fake news research. The problem took the form of a text classification task with six possible categories. Support Vector Machines (SVMs), Logistic Regression, Bi-directional Long Short-Term Memory (Bi-LSTMs), and Convolutional Neural Networks (CNNs) were used as reference models. Moreover, several combinations of text plus meta-data were used in CNN architectures to assess performance. Results showed that the statement's text, combined with all the meta-data yielded the best accuracy; however, this accuracy is barely 27.4%, which argues for the difficulty of the task at hand.

Another model trained and tested on the same dataset is multi-source multi-class fake news detection framework (MMFD) [8]. The architecture is based on deep learning techniques such as CNN and LSTM for feature extraction from the statement, an attention mechanism for combining information from multiple sources, and a function called multi-class discriminative function (MDF), whose purpose is to provide a better discriminative power between classes. Multiple inputs are used in the model: (a) the statements classified into six different classes, (b) meta-data consisting of a textual context about statements and speakers, (c) history, a vector that indicates a speaker's count of statements and (d) a report to argue that other information apart from the texts of the statements has the power to improve detection results. The report section contains verdict reports generated by experts in PolitiFact. Using only the text as feature input, the MMFD model obtained a mere 29.06% accuracy. Adding meta-data and history to the MMFD model increased the performance of the model to an accuracy of 34.77%. Moreover, adding the report source to the input of the model enhanced accuracy even more, up to 38.81%. The constant increase in the performance when supplementing the input with more sources shows that other information related to statements can improve the models' ability to better detect fake claims. This is the same argument that we are trying to build in our study, namely that the performance of classification models should increase by adding more context.

Oshikawa et al. [12] present a survey on fake news that exposes methods in which the problem of identifying fake news is addressed. The main highlighted categories are fact-checking, rumour detection, stance detection and sentiment analysis. The main categories in terms of sources for datasets consist of claims, entire articles and social networking services (SNS). Claims refer to short statements made by politicians or other public figures which are manually collected and labelled. The first dataset of this kind was constructed by Vlachos et al. [18], but it contained only 221 claims. Another dataset introduced by Fever [16] consisted of claims generated from Wikipedia data and provided related evidence for fact-checking. The entire articles category refers to datasets for predicting fake news where the article's entire text is labelled fake or true. One such example is FakeNewsNet [14] consiting of headlines and articles from PolitiFact and BuzzFeed. SNS are related to posts from social network, together with corresponding user profiles and links.

Another important analysis from the study of Oshikawa et al. points towards NLP and Machine Learning (ML) methods used to address the problem of fake news. Term Frequency-Inverse Document Frequency (TF-IDF), word2vec, or Global Vectors for Word Representation (GloVe) are commonly used to extract initial features. SVM, Naïve Bayes Classifiers, Logistic Regressions and Random Forests Classifiers are used as baselines in different studies. Classical text representation with neural networks, like LSTMs or CNNs, have also been used. The best results yielded by the study are: 41.5% accuracy for the LIAR dataset obtained by Kirilin and Strube [9] by replacing the credibility history with a larger credibility source, 64.7% accuracy achieved by Hanselowski et al. [6] for FEVER, and 94.4% accuracy obtained by Deligiannis et al. [3] on the BuzzFeed data from FakeNewsNet. Performance among datasets cannot be directly compared, since the LIAR dataset uses six possible labels, FEVER only three, and FakeNewsNet divides the data into fake or real. Newer methods for fake news identification [7, 15] consider deep learning techniques, more specifically Transformer-base language model [4], which is also considered in our work and is presented in detail in the method section.

16.3 Method

16.3.1 Corpus

Factual is a similar alternative to the already popular resource in fake news analysis, namely PolitiFact. *Factual* focuses on Romanian politics and highlights statements interpretations, as well as fact-checking relevant information. As with PolitiFact, the team of volunteers responsible for the selection and analysis of short statements is coordinated by experienced editors alongside experts in several fields, such as justice, politics, economics or public administration. Founded in 2014, *Factual* supplies interpretations for approximately 1000 statements. Particularly interesting when compared to Politifact is the wider range of labels assigned to the statements, containing custom additional nuances for specific situations: "adevarat" (eng., "true"), "partial adevarat" (eng., "partially true"), "numai cu sprijin institutional" (eng., "only with institutional support"), "trunchiat" (eng., "truncated"), "fals" (eng. "false"), "imposibil de verificat" (eng., "impossible to verify"), "in mandat" (eng., "during the mandate") and "in afara mandatului" (eng., "outside the mandate").

The interpretation of each analysed statement contains clear explanations, examples, and arguments for how and why a particular sentence was classified. Each statement is accompanied by a dedicated page where the team describes in detail the process behind validating the statements, references to articles of law or web pages containing additional information or data required to fully understand the context. We found this validation section particularly useful in our experiments, since the statements alone are not enough to classify the texts. Our objective is to show that having access to data that may confirm or deny a particular statement influences the computational interpretation of such texts. Additionally, statements are also classified in several categories such as finance, health, politics, which may prove beneficial for experiments that involve clustering or augmentations in specific fields.

The *Factual* website was crawled using the BeautifulSoup library,[3] a handy Python tool for extracting data from HTML Web pages. The crawling process was initiated from the main website, where the statements were grouped in short cards containing limited information. The dedicated page for each statement was then extracted, including the actual statement text, the previously mentioned validation section, the author, the label and the category. Due to the limited samples of a few of the custom labels (e.g., 5 statements for the "outside the mandate" category), we decided to perform a selection of the crawled data and only preserve the four main categories: true, mostly true, mostly false and false statements. The other entries were discarded from the corpus, which left us with a total of 845 statements, distributed as shown in Fig. 16.1.

We consider *Factual* a good starting point for high-quality annotated Romanian statements due to the rigorous underlying process for labelling entries, despite its small size.

16.3.2 Neural Architecture

State-of-the-art NLP models for text classification share the Transformer architecture [17] which surpasses the performance of older Recurrent Neural Network models; as such, the baseline for the *Factual* dataset consists of a similar model. Hug-

---

[3] https://pypi.org/project/beautifulsoup4/ Retrieved March 15, 2021.

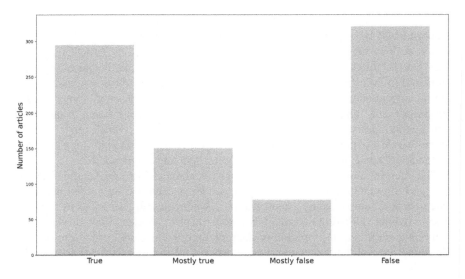

Fig. 16.1 Distribution of articles in the *Factual* dataset

gingface [22] is a well-known repository for Transformer models, including pre-trained multilingual and language-specific models. All follow-up experiments rely on RoBERT [11].

Bi-directional Encoder Representations from Transformers (BERT) models [4] can also be used for classification tasks with two input texts, like question answering, paraphrasing or summary evaluation. The two input texts are separated by a special token, which is the same as in the following sentence prediction pretraining task. In our case, the two texts are the statement and its corresponding context. However, the full contexts are too long to be used as input, since the RoBERT model was trained with a maximum length of 512 tokens, which should include both the statement and the context (see Table 16.1). Furthermore, given the limited number of examples in the dataset, it is unlikely for the model to learn to extract the relevant information from a very long text. Additionally, if the entire context is used, the statement and its context would differ significantly, with the context being considerably larger than the statement. However, our aim is to maintain the statements as central elements of the input and, given the length limitation imposed by BERT, our experiments were performed with only the first and the last paragraphs from the context.

Figure 16.2 features a block diagram showing the process in which the output of the "CLS" token from BERT was used for classification, by passing it through a dropout layer, a hidden layer with "tanh" activation and a final dense layer with softmax for the probabilities of the four classes. The model was initially trained by keeping the layers of the RoBERT model fixed and with a learning rate of $1e-3$; afterwards, the entire model was finetuned with a learning rate of $1e-5$.

Table 16.1 Statistics on statements and contexts

	Mean length (characters)	Mean length (words)
Statement	226	37
First paragraph of the context	224	34
Last paragraph of the context	184	28
Full context	1947	284

Fig. 16.2 BERT-based neural architecture

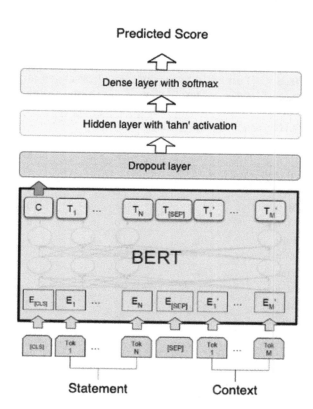

16.4 Results

From the initial dataset, 171 statements were kept for testing, while the rest of 674 were used for training and validation. Due to the limited size of the dataset, the model hyper-parameters were chosen using a fivefold cross-validation on the training partition. Early stopping was used for the both training steps (frozen RoBERT and finetuning).

The best models were evaluated on the test partition. The average number of epochs used in the cross-validation step was used for training the model on the entire training partition. Due to the high variability in the model's performance, a

Table 16.2 Model evaluation

RoBERT	Context	Dropout	Hidden	CV loss	CV Acc (%)	Test acc. (%)		
						Min.	Max.	Avg.
Base	No	0	32	1.3266	35.76	33.33	45.03	39.71
Base	Yes	0	32	1.2865	39.02	38.01	44.44	40.94
Base	First	0	32	1.2806	41.39	35.09	45.61	42.16
Base	Last	0	32	1.3772	35.75	35.67	43.27	40.23
Base	First	0.2	32	1.2608	40.94	35.67	44.44	40.64
Small	First	0.2	32	1.2976	39.17	32.16	45.61	36.78
Base	First	0.2	64	1.3284	39.16	38.01	47.37	42.63
Base	First	0.2	16	1.2864	39.46	31.58	45.03	40.41
Base (unw.)	First	0.2	32	**1.2430**	**44.07**	**43.86**	**49.71**	**46.84**

Bold denotes the best performing model

Table 16.3 Confusion matrices

	True	M. true	M. false	False
(a) *Unweighted loss*				
True	39	22	4	28
M. true	0	1	0	0
M. false	0	0	0	0
False	20	8	12	37
(b) *Weighted loss*				
True	24	13	1	13
M. true	11	9	2	12
M. false	8	2	9	11
False	16	7	4	29

consequence of the small size of the dataset, the training and testing were repeated for a number of ten times. The cross-validation results, as well as the minimum, maximum and average test accuracy are presented in Table 16.2.

All experiments except for the last model (Base unw.) use a class weighting factor proportional to the ratio between the number of examples in the class and the number of examples in the most common class. Although the best accuracies are obtained without the weighting factor (last line in the table), the model can only correctly predict the two ends of the spectrum (false and true), as seen in the confusion matrix given in Table 16.3a. The equivalent model trained with a weighted loss obtains a lower accuracy, but with more balanced predictions, as given in Table 16.3b. When comparing the average $F1$-scores for all labels, the unweighted model achieves a 27.42% accuracy, while the weighted version yields an accuracy of 39.60%.

16.5 Discussions

This sections addresses drawbacks of the devised method and describes potential improvements that may lead to new experiments. First, the size of the dataset is a major limitation, as advanced deep learning techniques require thousands of labelled entries; however, these were all the thoroughly evaluated statements available on the *Factual* website at this time.

Second, the high diversity in the topics covered by the entries topics and the underlying categories (see Table 16.4) creates an imbalance in the network training, especially when relating to the corresponding context—each dominant category (e.g., like politics, justice or the COVID-19 pandemic) contains more examples than all other salient categories with less than ten samples combined (e.g., agriculture, culture, sport or tourism).

Table 16.4 Distributions of samples per category

Factual category	Count
Politics	152
Justice	96
Financial	80
Electoral	80
Coronavirus	65
Economics	57
Health	41
Education	39
Social	38
Transport	35
External news	28
Europe	26
Work	26
Presidential	19
Environment	18
Energy	16
Business environment	9
Agriculture	5
Culture	4
Defense	3
Sport	3
Changes of mind	3
Tourism	1
Industry	1

Third, the initial *Factual* dataset contained particular labels such as: "with insti-tutional support only", "during the mandate" or "outside the mandate". Due to their personal note, their interpretability and the limited number of samples, we decided to exclude these statements and focus on truth-based labels for a clearer analysis. However, our dataset was further diminished by 10% because of this exclusion. Sim-ilar, but more established fact-checking initiatives like PolitiFact have a clearer list of labels, which are easier to separate, understand and interpret, and contain signif-icantly more entries (i.e. 20.000). As such, benchmarks like the LIAR dataset [21] allow for a wider range of experiments and debates throughout research groups due to the increased amount of clear separated statements. Nevertheless, we consider that initiatives like Factual are helpful for experiments in low-resource languages, such as Romanian.

Fourth, the two confusion matrices given in Table 16.3 provide a reflection on the suitability of devised categories when classifying fake news. Admittedly, the prediction accuracies for intermediate categories (mostly true and mostly false) are extremely low. As such, these results raise the question of whether the categories themselves should be further refined for an adequate classification and analysis of fake news. To further argue in favour of this point of view, we combine the subtypes distinguished through "mostly" as integral parts of the basic categories ("true" versus "false"); the prediction accuracy increases significantly, reaching 68.88% for "true" and 60.49% for "false", with an overall accuracy of 64.91%. This could indicate that RoBERT's performance allows it to fulfil its basic function (that of distinguish-ing between "true" statements and "false" statements), albeit being less efficient at supporting more elaborate explanations regarding the nature of the two classes. More-over, this could also mean that the attempt to classify news according to a gradual parameter can be deceptive and that, in its place, other more sophisticated typologies must be devised in addition to the fundamental dichotomy of "true" versus "false".

16.6 Conclusions and Future Work

We introduce a new dataset of fake news in Romanian consisting of statements and contexts with four different authenticity labels imported from a reliable source, the *Factual* project. Despite the limited number of entries available on the website, we consider that the rigorous analysis of political statements made throughout speeches, articles or other resources is a strong point of departure for performing NLP tech-niques on exclusively Romanian language data, while considering state-of-the-art Transformer models.

A series of experiments were performed, establishing a strong baseline that relies on the RoBERT model. Our results are comparable to the ones obtained by Wang et al. [21] on a similar dataset and demonstrate that the task at hand is extremely difficult. Moreover, this endeavour is a good starting point for the analysis of fake news in languages other than English.

As the Romanian statements from *Factual* present a similar structure and have similar labels as the popular initiative *PolitiFact*, a potential next step is to perform cross-lingual and transfer learning experiments using multilingual models trained on multiple datasets. Albeit the experiments show that adding context improves accuracy, the process of selecting the most relevant context is not addressed in this work. We plan to create a dataset from relevant online data sources for political statements and explore information retrieval methods for automatically selecting related sentences for a given statement.

Another potential step is that of exploring the possibility of extending the statement classification system beyond the "true" versus "false" dichotomy. In this regard, several combined approaches from the field of linguistics, analytic philosophy, and literary theory could prove their suitability in showing that not everything that is not factually "true" naturally becomes "false", but could rather fall under more complex categories such as "fictional", "imaginary", "subjective" or "ironic/satirical". An inquiry relying on a broader and more complex set of categories could significantly contribute to an increase in RoBERT's prediction accuracy.

Acknowledgements This work was supported by a grant of the Romanian Ministry of Education and Research, CNCS—UEFISCDI, project number PN-III-P1-1.1-TE-2019-1794, within PNCDI III. We would like to thank Ana Poenariu, the coordinator of the Factual project, for sharing the data and for her ongoing efforts to fight fake news in politics.

References

1. Allcott, H., Gentzkow, M.: Social media and fake news in the 2016 election. J. Econ. Perspect. **31**(2), 211–36 (2017)
2. Bovet, A., Makse, H.A.: Influence of fake news in twitter during the 2016 us presidential election. Nat. Commun. **10**(1), 1–14 (2019)
3. Deligiannis, N., Huu, T., Nguyen, D.M., Luo, X.: Deep learning for geolocating social media users and detecting fake news. In: NATO Workshop (2018)
4. Devlin, J., Chang, M.W., Lee, K., Toutanova, K.: BERT: Pre-training of deep bidirectional transformers for language understanding. In: Proceedings of the 2019 Conference of the North American Chapter of the Association for Computational Linguistics: Human Language Technologies, NAACL-HLT 2019, pp. 4171–4186 (2019)
5. Dizikes, P.: Study: On twitter, false news travels faster than true stories (2018). https://news.mit.edu/2018/study-twitter-false-news-travels-faster-true-stories-0308
6. Hanselowski, A., Zhang, H., Li, Z., Sorokin, D., Schiller, B., Schulz, C., Gurevych, I.: Ukpathene: Multi-sentence Textual Entailment for Claim Verification. arXiv:1809.01479 (2018)
7. Kaliyar, R.K., Goswami, A., Narang, P.: Fakebert: fake news detection in social media with a bert-based deep learning approach. Multimedia Tools Appl. **80**, 11765–11788 (2021)
8. Karimi, H., Roy, P., Saba-Sadiya, S., Tang, J.: Multi-source multi-class fake news detection. In: Proceedings of the 27th International Conference on Computational Linguistics, pp. 1546–1557. Association for Computational Linguistics, Santa Fe, New Mexico, USA (Aug 2018). https://www.aclweb.org/anthology/C18-1131
9. Kirilin, A., Strube, M.: Exploiting a speakers credibility to detect fake news. In: Proceedings of Data Science, Journalism and Media Workshop at KDD (DSJM18) (2018)

10. Lazer, D.M., Baum, M.A., Benkler, Y., Berinsky, A.J., Greenhill, K.M., Menczer, F., Metzger, M.J., Nyhan, B., Pennycook, G., Rothschild, D., et al.: The science of fake news. Science **359**(6380), 1094–1096 (2018)

11. Masala, M., Ruseti, S., Dascalu, M.: RoBERT—A Romanian BERT Model. In: Proceedings of the 28th International Conference on Computational Linguistics, pp. 6626–6637 (2020)

12. Oshikawa, R., Qian, J., Wang, W.Y.: A survey on natural language processing for fake news detection. arXiv:1811.00770 (2018)

13. Saikh, T., De, A., Ekbal, A., Bhattacharyya, P.: A deep learning approach for automatic detection of fake news. arXiv:2005.04938 (2020)

14. Shu, K., Mahudeswaran, D., Wang, S., Lee, D., Liu, H.: Fakenewsnet: a data repository with news content, social context, and spatiotemporal information for studying fake news on social media. Big Data **8**(3), 171–188 (2020)

15. Singhal, S., Shah, R.R., Chakraborty, T., Kumaraguru, P., Satoh, S.: Spotfake: A multi-modal framework for fake news detection. In: 2019 IEEE Fifth International Conference on Multimedia Big Data (BigMM), pp. 39–47. IEEE (2019)

16. Thorne, J., Vlachos, A., Christodoulopoulos, C., Mittal, A.: Fever: a large-scale dataset for fact extraction and verification. arXiv:1803.05355 (2018)

17. Vaswani, A., Shazeer, N., Parmar, N., Uszkoreit, J., Jones, L., Gomez, A.N., Kaiser, L., Polosukhin, I.: Attention is all you need. In: Advances in Neural Information Processing Systems, pp. 5998–6008 (2017). https://doi.org/10.1017/S0140525X16001837, http://papers.nips.cc/paper/7181-attention-is-all-you-need, http://arxiv.org/abs/1706.03762

18. Vlachos, A., Riedel, S.: Fact checking: task definition and dataset construction. In: Proceedings of the ACL 2014 Workshop on Language Technologies and Computational Social Science, pp. 18–22 (2014)

19. Vo, N., Lee, K.: The rise of guardians: fact-checking URL recommendation to combat fake news. In: The 41st International ACM SIGIR Conference on Research & Development in Information Retrieval, pp. 275–284 (2018)

20. Vosoughi, S., Roy, D., Aral, S.: The spread of true and false news online. Science **359**(6380), 1146–1151 (2018)

21. Wang, W.Y.: Liar, liar pants on fire: a new benchmark dataset for fake news detection. arXiv:1705.00648 (2017)

22. Wolf, T., Debut, L., Sanh, V., Chaumond, J., Delangue, C., Moi, A., Cistac, P., Rault, T., Louf, R., Funtowicz, M., Davison, J., Shleifer, S., von Platen, P., Ma, C., Jernite, Y., Plu, J., Xu, C., Scao, T.L., Gugger, S., Drame, M., Lhoest, Q., Rush, A.M.: Transformers: state-of-the-art natural language processing. In: Proceedings of the 2020 Conference on Empirical Methods in Natural Language Processing: System Demonstrations, pp. 38–45. Association for Computational Linguistics, Online (Oct 2020). https://www.aclweb.org/anthology/2020.emnlp-demos.6

Chapter 17
Exploring a Large Dataset of Educational Videos Using Object Detection Analysis

Eduard Cojocea and Traian Rebedea

Abstract Processing data and analyzing it are very important these days and have many applications in real life. In the case of very large video datasets, it became counterproductive, if not impossible, to visualize, analyze, and label the videos manually, due to their sheer number and length. Thus, automatically explore video datasets to understand what the videos are depicting, both for having a general view of the videos and a per video understanding of the activities. For example, if they contain sensitive images, violence and other such activities the videos could be flagged for manual checking. We turned our attention toward understanding educational videos, due to the great potential of improving the educational process by understanding written data (such as forums, chats, documents), audio, and video data. In this paper, we continue our work regarding an exploratory analysis of videos collected using YouTube-8 M, with the "school" keyword in their metadata. Thus, we attempt to detect the type of activity in a video based on the number of unique people detected and tracked, and on the objects detected by YOLOv3. This allows us to detect the educational activities in the dataset and to estimate the distribution of the activities in videos uploaded worldwide.

Keywords Video learning analytics · Exploratory data analysis · Educational videos · Object detection · People tracking · Educational activity detection · Deep learning

17.1 Introduction

The educational process is the most important activity humans need to plan and optimize in order to survive and thrive. In this digitalized era, we have unimaginable opportunities to improve education with the help of many digital tools: online courses and materials, remote classes, recorded and live lessons, online homework checkers,

E. Cojocea (✉) · T. Rebedea
University Politehnica of Bucharest, 313 Splaiul Independentei, Bucharest, Romania

T. Rebedea
e-mail: traian.rebedea@cs.pub.ro

© The Author(s), under exclusive license to Springer Nature Singapore Pte Ltd. 2022   213
Ó. Mealha et al. (eds.), *Ludic, Co-design and Tools Supporting Smart Learning Ecosystems and Smart Education*, Smart Innovation, Systems and Technologies 249,
https://doi.org/10.1007/978-981-16-3930-2_17

etc. All these increase the number and variety of ways in which information can reach people, representing a quantitative approach. But there is another important aspect to the equation: the quality of education. Evaluating the quality of a class, material, teacher, etc., is a difficult task, which has to be done with great care. Until recently, the vast majority of educational data available for analysis was in the written form: books, texts, online websites, forums, grading registers, etc. These days, the quantity of video recordings of the educational process has been increased drastically, due to a few factors. Firstly, the technology makes it possible for virtually everyone to have a camera in their pocket. Thus recording a class, a course, etc., is cheaper and easier than in the past. Secondly, people want more transparency in most societal activities, so there is an increasing pressure upon schools to video record their activities. Thirdly, the various online platforms that allow free uploading of videos make it easier for people to put together and organize educational videos without having to worry about maintaining such a platform.

Having a great amount of educational materials is great, but it comes with a problem. While there are many data mining [1, 2] methods for text-based data, exploring visual educational data is still in its early days. Due to the massive progress made in the last decade regarding image processing techniques using deep convolutional neural networks, these days are possible to automatically analyze videos and understand the activities happening in them, and used in conjunction with audio and text data we could analyze huge amounts of data in order to better understand and improve the educational process.

17.2 Related Work

In this section, we present existing methods regarding educational analytics and the networks we use for object detection and tracking. Presenting modern methods for processing videos allow us to see the full extent of possibilities regarding video analysis. Thus, we could explore and filter large datasets of videos using these tools are increasingly more available and more easy to use.

17.2.1 Educational Video Analytics

Relevant insights regarding the educational process can be extracted by using learning analytics (LA) and educational data mining (EDM). Then, these insights can be used to better understand and improve how we educate people. The vast majority of such solutions [3–5] focus on processing text-based data from massive open online courses, forums, social networks, etc.

Thus, despite some of the solutions proposed [6–9] hint at the importance of analyzing videos alongside text, most of them focus on manual labeling or processing video metadata, without taking visual information into consideration.

Some of the more recent studies propose solutions that take into consideration video and audio data. Ghauri et al. [10] propose a multimodal approach that uses audio, visual, and textual features to predict importance scores and how relevant are certain video segments. Another multimodal approach using audio and video features [11] is used to predict positive and negative climate in school classrooms videos.

We propose using modern computer vision techniques for detecting and tracking objects in videos, in order to extract insights from the video streams. Thus, we can find out if some objects of interest are present in a video, for example, sports-related gear (baseball bat, tennis racket, balls, etc.), and also how frequent such objects appear. Also, we can find a very good estimate of the number of unique people in the videos. Using this data, in conjunction we can realize two important analyses. First, we could group a large video dataset into clusters, for example, classroom videos, school commercials, canteen, indoor sports events, outdoor sports events, etc. Second, we could estimate a more specific activity taking place and extract a subset of videos with this activity, for example, chemistry classes (with relevant objects such as sinks and bottles, besides books, tables, chairs, and a relative low number of people), computer science classes (with relevant objects such as monitor, mouse and keyboard besides books, tables, chairs, and a relative low number of people).

17.2.2 Object Detection and Object Tracking

Computer vision is a field with many interesting tasks regarding processing visual data using heuristic or machine learning methods. Object detection is one of the most popular and most researched topics in this field, and it involves detecting, localizing, and classifying an object from an image. Most solutions these days use convolutional neural networks for such tasks.

There are many popular object detection solutions. A very popular series of solutions is region-based CNN (RCNN) [12], fast RCNN [13], faster RCNN [14], mask RCNN [15], which identify candidate regions which might contain objects and then run a classifier over each region.

Another popular family is You Only Look Once (YOLO), which are traversing the image only once, splitting the image into a grid, each cell of the grid is being responsible with identifying object in their region. Thus, YOLO [16], YOLO9000 [17], YOLOv3 [18], YOLOv4 [19] are generally faster than RCNN models and have comparable performance, which makes them a great choice for our purposes, since we have to process a big video dataset.

Object tracking is another very popular topic, which comes into two forms: tracking a single object and multiple object tracking. For our purposes, we need to track multiple objects, so we will not discuss the first case.

A naive solution for tracking objects is to use the Euclidian distance, but this solution struggles to track correctly even in images with few objects. A good improvement is to use a Kalman filter [20], as presented in simple online and real-time tracking (SORT) [21]. Further, deep SORT [22] improves the tracking performance for the particular case of people tracking, since it uses a deeply associated metric learned using a pretrained CNN on a pedestrian dataset [23].

17.3 Proposed Method

17.3.1 Dataset

The same as in our previous paper, we use YouTube-8M [24] to retrieve a dataset of educational videos. Thus, from the 7815 videos with the tag "school," we extract 1912 videos which met the quality requirements and were relevant for our research.

Since some videos are quite large, we extract a roughly two-minute section of each video, taken from the middle, in order to avoid intro and outro scenes, thus getting relevant footage. Furthermore, this helps with the analysis, since all videos will be of a similar length. The same as in our previous paper, this is done to keep the computational cost reasonable regarding the time for processing.

17.3.2 People Detection and Tracking

As presented in our previous paper [25], we have the results regarding the unique number of people in each video, based on YOLOv3 [18] object detection and Deep-SORT [20] people tracking. Thus, we will use the results from our previous paper in conjunction with other detections in order to better understand and group the videos.

17.3.3 Object Detection

YOLOv3 can be used to detect 80 objects, some of which are relevant for school environments. We used YOLOv3 to detect objects in all the videos and took into consideration only a subset of objects which we considered relevant. We have done this to keep the noise level in our detections to a minimal, since some objects are very unlikely to be present in such environments: aero plane, giraffe, bear, etc., and they could be accidental detections or detections in printed formats (books, papers, backpacks, etc.), which will falsely signal that a bear is in a school. The objects that we took into consideration and group by handmade categories are presented in Fig. 17.1.

It is important to mention that the tracking was done solely for persons. When it comes to objects, the number of detections does not reflect the unique number of those

```
{
    "transport": ["bicycle", "car", "motorbike", "bus"],
    "food": ["banana", "apple", "sandwich", "orange","broccoli", "carrot", "hot dog", "pizza", "donut", "cake"],
    "eating": ["bottle", "cup", "fork", "knife", "spoon", "bowl"],
    "furniture": ["chair", "sofa", "diningtable", "bench"],
    "sports": ["sports ball", "baseball bat", "baseball glove", "tennis racket"],
    "luggage": ["backpack", "handbag", "suitcase", "umbrella", "tie"],
    "study_objects": ["tvmonitor", "laptop", "mouse", "remote", "keyboard", "cell phone", "clock", "book", "scissors", "teddy bear"],
    "wc": ["toilet", "sink"]
}
```

Fig. 17.1 Relevant objects detected by YOLOv3 grouped by categories

objects, but merely how many times YOLOv3 has detected such an object during the entire video. For example, if YOLOv3 detects thousands of "TV monitors" in a video, it could mean that we have tens, hundreds, or thousands of such objects, but it may as well mean that there is one single such object (like a blackboard, electronic board, etc.) which is visible in most frames.

Another important thing to mention is that, due to the lack of tracking, accidental erroneous detections may occur and introduce noise into our analysis. For example, a certain object could be detected in a few frames, but we will ignore such detections, considering that relevant objects will be visible for at least 1 s, which means that we would expect at least 20–24 detections. Thus, any object with lower detections that this will not be considered.

17.3.4   Clustering

We use K-means and mean-shift clustering methods in order to detect groups of similar videos regarding activities. We cluster both one-dimensional arrays of persons, objects, and categories and two-dimensional combinations of them.

17.4   Results

We first analyze the videos based solely on the objects detected, excluding people, and then we combine the findings from the previous paper with the ones in this paper.

17.4.1   Grouping by the Dominant Object

Our first step in exploring the results was to see how the videos distributions look like based on the most detected object. The results presented in Fig. 17.2 show us some insights regarding the videos. Unsurprisingly, most videos have the dominant objects "chair" and "book." This is to be expected from school videos, and the fact

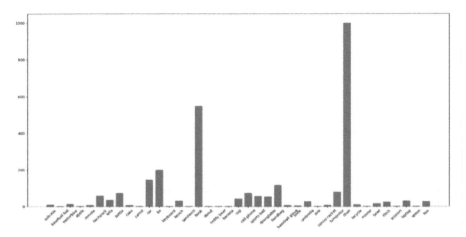

Fig. 17.2 Distribution of the dominant object in the videos

that they make more than half of the dataset is a good indicator for school-related videos. Then, we have another significant amount of videos with the "tie" as the dominant object, which indicates the existence of school uniforms. Surprisingly, there are many videos where the object "car" is the dominant object, which could mean either students being driven to school by parents, or that the video is being taken in the school parking lot. Then we have sports-related school activities, where the dominant objects are "sports ball," "tennis racket," "baseball bat," etc., which are clear indicators of sports activities happening in the video. Other videos with the dominant objects such as food items, dining tables, and cups indicate a scene from a school canteen.

17.4.2 Grouping by the Top Three Objects

In this subsection, we describe the distribution of the videos based on the top three detections of objects. In our analysis, we did not take into considerations which of the objects are first, second or third, just that they are in the top three. For example, a video with the top three objects "book," "chair," "bottle" has the same top three with a video with the top three objects "chair," "bottle," "book."

Thus we created new labels, which contain the first four letters of each object, sorted alphabetically, concatenated, and separated by underscore characters. This is done for readability purposes.

The results presented in Fig. 17.3 provide deeper insights regarding the activities in the videos. The most common top three is composed of the objects "car," "chair," and "dining table." It is important to remind that the order is alphabetical. This is somewhat surprising, since we would expect this top three for a restaurant, but we

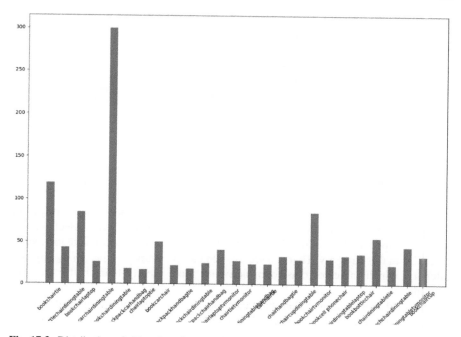

Fig. 17.3 Distribution of videos based on top three object detections

should keep in mind that many of the videos in our dataset are school commercials which could show parking lots, classrooms, the canteen, etc., which could justify our results.

The second most common top three is "book," "chair," "tie" which indicates scenes from a classroom of students in uniform reading or writing. Similar activities are depicted by the third most common top three: "book," "chair," "laptop," but done in science classrooms.

The fourth most common top three is "chair," "cup," "dining table," which indicate scenes from the school canteen of cafeteria, the existence of which is a highly valued feature by parents and children.

17.4.3 Grouping by the Dominant Category

The categories are presented in Sect. 17.3.3, Fig. 17.1. They represent a grouping of objects by their purpose done by us in order to get better insights. For example, all food items ("banana," "sandwich," "pizza," etc.) are grouped in the "food" category, cutlery and other eating-related objects ("bottle," "knife," "bowl," etc.) are grouped under the "eating" category, and so on. Thus, situations could be avoided where the dominant objects in different videos are "banana," "apple," "sandwich" which all

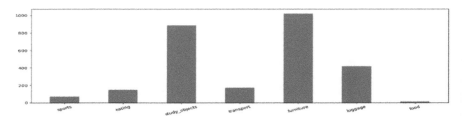

Fig. 17.4 Distribution of videos based on the dominant object category

indicate people eating could be lost in the histogram due to the fact that they are different objects.

The results presented in Fig. 17.4 show us how the video distribution looks like based on the dominant category of objects. Based on our findings in Sect. 17.4.1, we could expect that "furniture" will be the most common dominant category, since "chair" is in this category. Then we see that almost as common is the "study objects" category, which indicates classroom activities more clearly than in the previous sections.

The third most common dominant category is "luggage," which indicates either recess or commercials depicting children with their backpacks coming to or coming from classes.

The fourth most common dominant category is "transport" which indicates scenes from the parking lot, with parents driving their children or the school bus bringing students to class.

Then, we have the "eating" and "food" categories indicating canteen or cafeteria scenes. The first category is more common, because the latter contains only a small subset of possible food, while the cutlery is similar across the world.

And we have the "sports" category which clearly indicate sports activities based on sports-related objects "tennis racket," "sports ball," etc.

17.4.4 Grouping by the Top Three Categories

Similar to the analysis in Sect. 17.4.2, we group the videos based on the top three categories of objects detected. We create a label, based on the same rules, for readability purpose.

The results presented in Fig. 17.5 show us that by far the most common top three categories are "luggage," "study objects," "transport." The "transport" category is perhaps surprising, but the other two are to be expected. Then, we see that the "eating" and "food" categories are present in the next most common top three3 categories, alongside with "furniture" and "study objects."

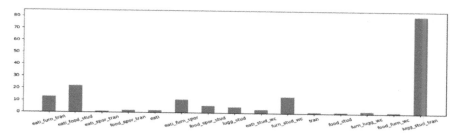

Fig. 17.5 Distribution of videos based on the top three object categories

17.4.5 Clustering Videos by Chair Detections

In Sect. 17.4.1, we observed that the "chair" object was by far the most common dominant object in the videos. Thus, we use the number of chair detections in the videos to cluster the videos, similar to what we did in our previous paper [25] with the number of persons. We excluded the videos where the number of detections were higher than 10 thousand, since we considered them as anomalies. Similarly, we excluded the videos with less than 24 detections, considering that an object should be visible at least for 1 s, resulting in at least 20–24 detections. Thus, a lower number could indicate erroneous detections.

By using mean-shift clustering on this data, we get two clusters, as presented in Fig. 17.6. Our interpretation of these clusters is that the first cluster, with less than 3750 chair detections generally represents videos with classroom or cafeteria activities, where chairs are visible, but are not in a great amount. The second cluster generally represents sports and music activities with chairs available for spectators.

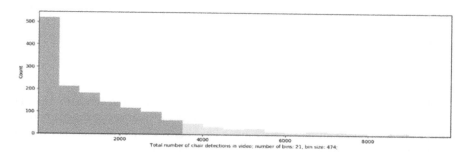

Fig. 17.6 Distribution of videos based on the number of chair detections. By using mean shift, there are two clusters detected

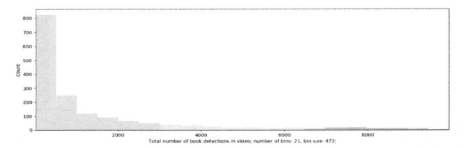

Fig. 17.7 Distribution of videos based on the number of book detections. By using mean shift, there are three clusters detected

17.4.6 Clustering by Book Detections

In Sect. 17.4.1, we observed that the second most detected object was "book." Thus, we use the number of book detections in the videos as well to cluster the videos; similarly as in the previous section, we excluded the videos where the number of detections where higher than 10 thousand, since we considered them as anomalies and the ones with less than 24 detections.

By using mean-shift clustering on this data, we get three clusters, as presented in Fig. 17.7. The first cluster with less than 3000 book detections generally represent videos with classroom activities, where books are visible, but are not the main focus. The second cluster generally represents school commercials, where the school library and book materials are presented, thus being among the objects present during most of the duration of the video. The third cluster with more than 6500 book detection generally contains videos with a single person presenting school equipment, among which there are many books and notebooks visible during the vast majority of frames.

17.4.7 2D Clustering

Our previous paper was pretty straightforward when extracting insights from visual data, because it relied on a single visual information: the number of unique people in a video. In the current paper, we have many more relevant objects, and thus a complete thorough analysis of how these objects detected can be used to understand the video dataset is beyond the scope of the paper. Nevertheless, we will present some results obtained by 2D clustering the videos based on: book–chair, person–book, person–chair pairs.

In Figure 17.8 are presented 10 clusters detected using mean shift applied on book–chair data; in Fig. 17.9, five clusters for person–book data; and in Fig. 17.10, three clusters for person–chair data. This indicates many possible subcategories of activities in videos. Thus, by using an object detector which could detect a significant

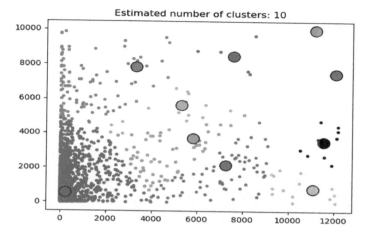

Fig. 17.8 On the *x*-axis are the number of book detections and on the *y*-axis are the number of chair detections. Ten clusters detected using mean shift

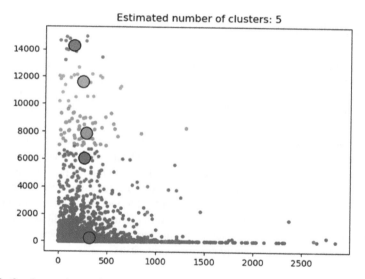

Fig. 17.9 On the *x*-axis are the number of persons and on the *y*-axis are the number of book detections. Five clusters detected using mean shift

number of important educational-related objects and clustering data combining these detections, we could obtain finer and finer categories of activities.

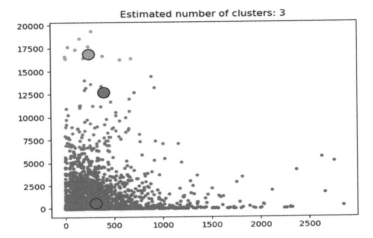

Fig. 17.10 On the x-axis are the number of persons and on the y-axis are the number of chair detections. Three clusters detected using mean shift

17.5 Conclusions

In conclusion, we are aware that the analysis in this paper is only incipient; but by using our proposed solution, we could explore large datasets of videos and understand what kind of activities are taking place. Thus, we can filter, sort, and group such videos, making it easier to apply further evaluation techniques for the videos from a category of interest.

Furthermore, this paper expands upon the findings of our previous paper, showing that we can use the number of people in conjunction with other objects detected in order to better understand the visual information. Thus, a person interested in analyzing specific school activities like computer science classes, chemistry classes, and specific sports activities could just search for the videos which contain the relevant object with a reasonable number of detections. This will significantly reduce the time spent manually evaluating videos, by having a person to process only a small subset of videos, which are candidates for its interests.

References

1. Dutt, A., Ismail, M.A., Herawan, T.: A systematic review on educational data mining. IEEE Access **5**, 15991–16005 (2017)
2. Ferguson, R.: Learning analytics: drivers, developments and challenges. Int. J. Technol. Enhanced Learn. **4**(5/6), 304–317 (2012)
3. Kennedy, G., Coffrin, C., Barba, P.D., Corrin, L.: Predicting success: how learners' prior knowledge, skills and activities predict MOOC performance. In: Proceedings of the Fifth International Conference on Learning Analytics and Knowledge, pp. 136–140 (2015)

4. Sabourin, J., Kosturko, L., FitzGerald, C., McQuiggan, S.: Student privacy and educational data mining: perspectives from industry. In: International Conference on Educational Data Mining Society (2015)
5. Trausan-Matu, S., Dascalu, M., Rebedea, T.: PolyCAFe—automatic support for the polyphonic analysis of CSCL chats. Int. J. Comput.-Support. Collab. Learn. 9(2), 127–156 (2014)
6. Chatbri, H., Oliveira, M., McGuinness, K., Little, S., Kameyama, K., Kwan, P., O'Connor, N.E.: Educational video classification by using a transcript to image transform and supervised learning. In: 2017 Seventh International Conference on Image Processing Theory, Tools and Applications (IPTA). IEEE (2017)
7. Li, X., Wang, M., Zeng, W., Lu, W.: A students' action recognition database in smart classroom. In: 2019 14th International Conference on Computer Science and Education (ICCSE), pp. 523–527. IEEE (2019)
8. Radloff, J., Guzey, S.: Investigating changes in preservice teachers' conceptions of STEM education following video analysis and reflection. Sch. Sci. Math. 117(3–4), 158–167 (2017)
9. Shoufan, A.: Estimating the cognitive value of YouTube's educational videos: a learning analytics approach. Comput. Hum. Behav. 92, 450–458 (2019)
10. Ghauri, J. A., Hakimov, S., Ewerth, R. Classification of important segments in educational videos using multimodal features. arXiv Preprint arXiv:2010.13626 (2020)
11. Ramakrishnan, A., Zylich, B., Ottmar, E., LoCasale-Crouch, J., Whitehill, J.: Toward automated classroom observation: multimodal machine learning to estimate class positive climate and negative climate. IEEE Trans. Affective Comput. (2021)
12. Girshick, R., Donahue, J., Darrell, T., Malik, J.: Rich feature hierarchies for accurate object detection and semantic segmentation. In: Proceedings of the IEEE Conference on Computer Vision and Pattern Recognition, pp. 580–587 (2014)
13. Girshick, R.: Fast R-CNN. In: Proceedings of the IEEE International Conference on Computer Vision, pp. 1440–1448 (2015)
14. Ren, S., He, K., Girshick, R., Sun, J.: Faster R-CNN: towards real-time object detection with region proposal networks. In: Advances in Neural Information Processing Systems, pp. 91–99 (2015)
15. He, K., Gkioxari, G., Dollár, P., Girshick, R.: Mask R-CNN. In: Proceedings of the IEEE International Conference on Computer Vision, pp. 2961–2969 (2017)
16. Redmon, J., Divvala, S., Girshick, R., Farhadi, A.: You only look once: unified, real-time object detection. In: Proceedings of the IEEE Conference on Computer Vision and Pattern Recognition, pp. 779–788 (2016)
17. Redmon, J., Farhadi, A.: YOLO9000: better, faster, stronger. In: Proceedings of the IEEE Conference on Computer Vision and Pattern Recognition, pp. 7263–7271 (2017)
18. Redmon, J., Farhadi, A.: Yolov3: an incremental improvement. arXiv Preprint arXiv:1804.02767 (2018)
19. Bochkovskiy, A., Wang, C.Y., Liao, H.Y.M.: Yolov4: optimal speed and accuracy of object detection. arXiv Preprint arXiv:2004.10934 (2020)
20. Welch, G., Bishop, G.: An introduction to the Kalman filter (1995)
21. Bewley, A., Ge, Z., Ott, L., Ramos, F., Upcroft, B.: Simple online and realtime tracking. In 2016 IEEE International Conference on Image Processing (ICIP), pp. 3464–3468, IEEE, (2016)
22. Wojke, N., Bewley, A., Paulus, D.: Simple online and realtime tracking with a deep association metric. In: 2017 IEEE International Conference on Image Processing (ICIP), pp. 3645–3649. IEEE (2017)
23. Zheng, L., Bie, Z., Sun, Y., Wang, J., Su, C., Wang, S., Tian, Q.: Mars: a video benchmark for large-scale person re-identification. In: European Conference on Computer Vision, pp. 868–884. Springer, Cham (2016)
24. Abu-El-Haija, S., Kothari, N., Lee, J., Natsev, P., Toderici, G., Varadarajan, B., Vijayanarasimhan, S.: Youtube-8m: a large-scale video classification benchmark. arXiv Preprint arXiv:1609.08675 (2016)
25. Cojocea, E., Rebedea, T.: Exploratory analysis of a large dataset of educational videos: preliminary results using people tracking. In: Ludic, Co-design and Tools Supporting Smart Learning Ecosystems and Smart Education, pp. 211–223. Springer, Singapore (2020)

Correction to: Robots as My Future Colleagues: Changing Attitudes Toward Collaborative Robots by Means of Experience-Based Workshops

Janika Leoste, Tõnu Viik, José San Martín López, Mihkel Kangur,
Veiko Vunder, Yoan Mollard, Tiia Õun, Henri Tammo, and Kristian Paekivi

Correction to:
Chapter 13 in: Ó. Mealha et al. (eds.), *Ludic, Co-design*
and Tools Supporting Smart Learning Ecosystems and Smart
Education, **Smart Innovation, Systems and Technologies 249,**
https://doi.org/10.1007/978-981-16-3930-2_13

The original version of the chapter "Robots as My Future Colleagues: Changing Attitudes Toward Collaborative Robots by Means of Experience-Based Workshops" without open access. This has now been changed to the copyright holder "The Author(s)" and open access under the terms of the Creative Commons Attribution 4.0 International License (https://creativecommons.org/licenses/by/4.0/). For further details see license information in the chapter.

The chapter and the book have been updated with the change.

The updated version of this chapter can be found at
https://doi.org/10.1007/978-981-16-3930-2_13

Author Index

Ó. Mealha et al. (eds.), *Ludic, Co-design and Tools Supporting Smart Learning Ecosystems and Smart Education*, Smart Innovation, Systems and Technologies 249,
https://doi.org/10.1007/978-981-16-3930-2

Lightning Source UK Ltd.
Milton Keynes UK
UKHW020605050922
408354UK00002B/28

9 789811 639326